Ben Stacy Jerrik (Ed.)

Camorino

Ben Stacy Jerrik (Ed.)

Camorino

Cantons of Switzerland, Bellinzona (district), Communes of Switzerland

Part Press

Imprint

All parts of this book are extracted from Wikipedia, the free encyclopedia (www.wikipedia.org).

You can get detailed informations about the authors of this collection of articles at the end of this book. The editors (Ed.) of this book are no authors. They have not modified or extended the original texts.

Pictures published in this book can be under different licences than the GNU Free Documentation License. You can get detailed informations about the authors and licences of pictures at the end of this book.

Cover image: www.ingimage.com
Concerning the licence of the cover image please contact ingimage.

Publisher:
Part Press is a trademark of
International Book Market Service Ltd., 17 Rue Meldrum, Beau Bassin, 1713-01 Mauritius
Email: info@bookmarketservice.com
Website: www.bookmarketservice.com

Published in 2013

Printed in: U.S.A., U.K., Germany. This book was not produced in Mauritius.

ISBN: 978-620-1-55130-5

Contents

Articles

References

Camorino

Camorino	
Country	Switzerland
Canton	Ticino
District	Bellinzona
Population	2697 (Dec 2011)[1]
- Density	326 /km^2 (845 /sq mi)
Area	km^2 (sq mi)[2]
Elevation	223 m (ft)
Postal code	6528
SFOS number	5004
Surrounded by	Cadenazzo, Giubiasco, Isone, Medeglia, Pianezzo, Sant'Antonino
Website	www.camorino.ch [3] Profile [4] (**Italian**), SFSO statistics [5]

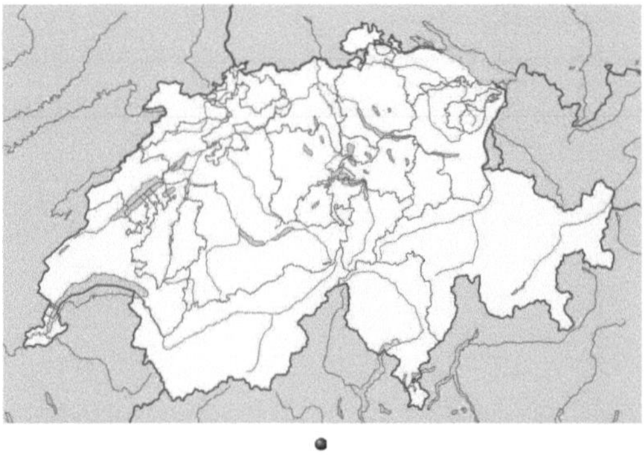

Camorino

Camorino is a municipality in the district of Bellinzona in the canton of Ticino in Switzerland.

History

From time immemorial the inhabitants of Isone regularly crossed the Tiglio alpine pass to arrive in the Camorino area, where they owned vineyards and arable land. Over the centuries, many of them finally created a village. Camorino is first mentioned in 1237 as part of the lands of the collegiate church of S. Pietro in Bellinzona.[6] A church is mentioned in the village in 1237 and 1285, but no visible traces have been preserved. The existing parish church of San Martino was built in 1553 and completely renovated in 1888. In 1583, the village church split from the Bellinzona parish and was confirmed in 1591 as a separate parish.[6]

In 1853 a defensive line between Sementina and Camorino was built to create jobs for the citizen of Ticino expelled from the Kingdom of Lombardy–Venetia. The *Fortini della fame*, five round towers with arrow slits in the upper part, is part of these fortifications.

Camorino was originally a pure farming village (raising mostly grapes, strawberries and livestock) with some handicraft, trade and light industry (manufacturing knitwear and pasta as well as a mill). Due to the lack of jobs, some of whose inhabitants emigrated to America. The population growth coincides with the growth of the cities of Giubiasco and Bellinzona. Today it is a residential community with many commuters, with no large industrial companies and with very little agriculture.[6]

Geography

Camorino has an area, as of 1997, of 8.28 square kilometers (sq mi). Of this area, 1.93 km^2 (sq mi) or 23.3% is used for agricultural purposes, while 5.55 km^2 (sq mi) or 67.0% is forested. Of the rest of the land, 1.05 km^2 (sq mi) or 12.7% is settled (buildings or roads), 0.05 km^2 (acres) or 0.6% is either rivers or lakes and 0.15 km^2 (acres) or 1.8% is unproductive land.

Of the built up area, housing and buildings made up 6.5% and transportation infrastructure made up 4.3%. Out of the forested land, 63.9% of the total land area is heavily forested and 2.8% is covered with orchards or small clusters of trees. Of the agricultural land, 11.4% is used for growing crops, while 5.4% is used for orchards or vine crops and 6.5% is used for alpine pastures. All the water in the municipality is flowing water. Of the unproductive areas, 1.7% is unproductive vegetation and .[7]

The municipality is located in the Bellinzona district, and is a suburb of Bellinzona. The village is crossed by several major roads and rail lines leading into Bellinzona.

Coat of arms

The blazon of the municipal coat of arms is *Gules on a Base Argent wavy two bars wavy Azure and overall a Mill Wheel Or.* The mill wheel refers to the *Maglio*, a mill which, with another house, remained undamaged after a landslide that destroyed the rest of the village.[8] "

Demographics

Camorino has a population (as of December 2011) of 2697.[1] As of 2008, 16.1% of the population are foreign nationals.[9] Over the last 10 years (1997–2007) the population has changed at a rate of 17.7%. Most of the population (as of 2000) speaks Italian(89.6%), with German being second most common (5.3%) and French being third (1.2%).[10]

House in Camorino in 1932

Of the Swiss national languages (as of 2000), 117 speak German 26 people speak French, 1,981 people speak Italian, and 2 people speak Romansh. The remainder (84 people) speak another language.[11]

As of 2008, the gender distribution of the population was 50.4% male and 49.6% female. The population was made up of 1,088 Swiss men (40.2% of the population), and 275 (10.2%) non-Swiss men. There were 1,139 Swiss women (42.1%), and 204 (7.5%) non-Swiss women.[12]

In 2008 there were 27 live births to Swiss citizens and 1 birth to non-Swiss citizens, and in same time span there were 24 deaths of Swiss citizens and 3 non-Swiss citizen deaths. Ignoring immigration and emigration, the population of Swiss citizens increased by 3 while the foreign population decreased by 2. There was 1 Swiss man and 1 Swiss woman who emigrated from Switzerland to another country, 9 non-Swiss men who emigrated from Switzerland to another country and 5 non-Swiss women who emigrated from Switzerland to another country. The total Swiss population change in 2008 (from all sources) was an increase of 26 and the non-Swiss population change was an increase of 19 people. This represents a population growth rate of 1.8%.[9]

The age distribution, as of 2009, in Camorino is; 303 children or 11.2% of the population are between 0 and 9 years old and 272 teenagers or 10.1% are between 10 and 19. Of the adult population, 316 people or 11.7% of the population are between 20 and 29 years old. 383 people or 14.2% are between 30 and 39, 481 people or 17.8% are between 40 and 49, and 370 people or 13.7% are between 50 and 59. The senior population distribution is 286 people or 10.6% of the population are between 60 and 69 years old, 179 people or 6.6% are between 70 and 79, there are 116 people or 4.3% who are over 80.[12]

As of 2000, there were 934 private households in the municipality, and an average of 2.4 persons per household.[10] In 2000 there were 482 single family homes (or 71.1% of the total) out of a total of 678 inhabited buildings. There were 120 two family buildings (17.7%) and 45 multi-family buildings (6.6%). There were also 31 buildings in the municipality that were multipurpose buildings (used for both housing and commercial or another purpose).[13]

The vacancy rate for the municipality, in 2008, was 0.08%. In 2000 there were 1,079 apartments in the municipality. The most common apartment size was the 4 room apartment of which there were 363. There were 53 single room apartments and 270 apartments with five or more rooms.[14] Of these apartments, a total of 928 apartments (86.0% of the total) were permanently occupied, while 126 apartments (11.7%) were seasonally occupied and 25 apartments (2.3%) were empty.[14] As of 2007, the construction rate of new housing units was 4.7 new units per 1000 residents.[10]

The historical population is given in the following table:[6]

year	population
1591	400
1698	211
1784	300
1850	321
1880	483
1900	405
1950	702
1980	1,476
1990	1,709
2000	2,210

Heritage sites of national significance

The Ottocentesche fortification and the *Fortini Della Fame* are listed as Swiss heritage sites of national significance.[15]

Politics

In the 2007 federal election the most popular party was the FDP which received 28.1% of the vote. The next three most popular parties were the CVP (24.34%), the SP (21.57%) and the Ticino League (12.74%). In the federal election, a total of 922 votes were cast, and the voter turnout was 54.3%.[16]

In the 2007 *Gran Consiglio* election, there were a total of 1,701 registered voters in Camorino, of which 1,239 or 72.8% voted. 14 blank ballots and 2 null ballots were cast, leaving 1,223 valid ballots in the election. The most popular party was the PLRT which received 290 or 23.7% of the vote. The next three most popular parties were; the PPD+GenGiova (with 253 or 20.7%), the PS (with 233 or 19.1%) and the SSI (with 232 or 19.0%).[17]

In the 2007 *Consiglio di Stato* election, there were 18 blank ballots and 2 null ballots, which left 1,219 valid ballots in the election. The most popular party was the PLRT which received 273 or 22.4% of the vote. The next three most popular parties were; the PS (with 268 or 22.0%), the PPD (with 246 or 20.2%) and the SSI (with 207 or 17.0%).[17]

Economy

As of 2007, Camorino had an unemployment rate of 4.7%. As of 2005, there were 71 people employed in the primary economic sector and about 19 businesses involved in this sector. 217 people are employed in the secondary sector and there are 23 businesses in this sector. 666 people are employed in the tertiary sector, with 80 businesses in this sector.[10]

There were 962 residents of the municipality who were employed in some capacity, of which females made up 40.2% of the workforce. In 2000, there were 660 workers who commuted into the municipality and 747 workers who commuted away. The municipality is a net exporter of workers, with about 1.1 workers leaving the municipality for every one entering. About 3.6% of the workforce coming into Camorino are coming from outside Switzerland.[18]

Of the working population, 8% used public transportation to get to work, and 67.8% used a private car.[10]

As of 2009, there was one hotel in Camorino.[19]

Religion

From the 2000 census, 1,776 or 80.4% were Roman Catholic, while 118 or 5.3% belonged to the Swiss Reformed Church. There are 246 individuals (or about 11.13% of the population) who belong to another church (not listed on the census), and 70 individuals (or about 3.17% of the population) did not answer the question.[11]

Village church in 1932

Education

The entire Swiss population is generally well educated. In Camorino about 70.6% of the population (between age 25-64) have completed either non-mandatory upper secondary education or additional higher education (either University or a *Fachhochschule*).[10]

In Camorino there are a total of 469 students (as of 2009). The Ticino education system provides up to three years of non-mandatory kindergarten and in Camorino there are 73 children in kindergarten.

The primary school program lasts for five years and includes both a standard school and a special school. In the municipality, 160 students attend the standard primary schools and 5 students attend the special school. In the lower secondary school system, students either attend a two-year middle school followed by a two-year pre-apprenticeship or they attend a four-year program to prepare for higher education. There are 113 students in the two-year middle school and in their pre-apprenticeship, while 44 students are in the four-year advanced program.

The upper secondary school includes several options, but at the end of the upper secondary program, a student will be prepared to enter a trade or to continue on to a university or college. In Ticino, vocational students may either attend school while working on their internship or apprenticeship (which takes three or four years) or may attend school followed by an internship or apprenticeship (which takes one year as a full-time student or one and a half to two years as a part-time student).[20]

There are 22 vocational students who are attending school full-time and 46 who attend part-time.The professional program lasts three years and prepares a student for a job in engineering, nursing, computer science, business, tourism and similar fields. There are 6 students in the professional program.[21]

As of 2000, there were 3 students in Camorino who came from another municipality, while 246 residents attended schools outside the municipality.[18]

References

[1] Swiss Federal Statistics Office – STAT-TAB (http://www.pxweb.bfs.admin.ch/Dialog/varval.asp?ma=px-d-01-2A01&ti=Ständige+und+Nichtständige+Wohnbevölkerung+nach+Region,+Geschlecht,+Nationalität+und++Alter&path=../Database/German_01 - Bevölkerung/01.2 - Bevölkerungsstand und -bewegung/&lang=1&prod=01&openChild=true&secprod=2) *Ständige und Nichtständige Wohnbevölkerung nach Region, Geschlecht, Nationalität und Alter* (German) accessed 3 October 2012

[2] Arealstatistik Standard - Gemeindedaten nach 4 Hauptbereichen (http://www.bfs.admin.ch/bfs/portal/de/index/themen/02/03/blank/data/gemeindedaten.html)

[3] http://www.camorino.ch

[4] http://www.ti.ch/DFE/USTAT/DATI_COMUNI/0_pagine/default.asp?numero=5004

[5] http://www.bfs.admin.ch/bfs/portal/en/index/regionen/02/key.html

[6] *Camorino* in German (http://www.hls-dhs-dss.ch/textes/d/D2033.php), French (http://www.hls-dhs-dss.ch/textes/f/F2033.php) and Italian (http://www.hls-dhs-dss.ch/textes/i/I2033.php) in the online *Historical Dictionary of Switzerland*.

[7] Altitudine, superficie, secondo il genere di utilizzazione, rilevazione 1992/1997, e densità della popolazione, nel 2000 (http://www.ti.ch/DFE/USTAT/DATI_RIEPILOGHI/default_x.asp?menu=02&key1_testo=Territorio e ambiente&key1=02&key2=01&key3=01) (Italian) accessed 25 October 2010

[8] Flags of the World.com (http://www.fotw.us/flags/ch-ti004.html) accessed 26-October-2010

[9] Swiss Federal Statistical Office - Superweb database - Gemeinde Statistics 1981-2008 (http://www.bfs.admin.ch/bfs/portal/de/index/infothek/onlinedb/superweb/login.html) (German) accessed 19 June 2010

[10] Swiss Federal Statistical Office (http://www.bfs.admin.ch/bfs/portal/en/index/regionen/02/key.html) accessed 26-October-2010

[11] Popolazione residente, secondo la lingua principale e la religione, nel 2000 (http://www.ti.ch/DFE/USTAT/DATI_RIEPILOGHI/default_x.asp?menu=01&key1_testo=Popolazione&key1=01&key2=05&key3=00) (Italian) accessed 23 November 2010

[12] 01.02.03 Popolazione residente permanente (http://www.ti.ch/DFE/USTAT/DATI_RIEPILOGHI/default_x.asp?menu=01&key1_testo=Popolazione&key1=01&key2=02&key3=03) (Italian) accessed 23 November 2010

[13] 09.02.01 Edifici (http://www.ti.ch/DFE/USTAT/DATI_RIEPILOGHI/default_x.asp?menu=09&key1_testo=Costruzioni e abitazioni&key1=09&key2=02&key3=01) (Italian) accessed 23 November 2010

[14] 09.02.02 Abitazioni (http://www.ti.ch/DFE/USTAT/DATI_RIEPILOGHI/default_x.asp?menu=09&key1_testo=Costruzioni e abitazioni&key1=09&key2=02&key3=02) (Italian) accessed 23 November 2010

[16] Swiss Federal Statistical Office, *Nationalratswahlen 2007: Stärke der Parteien und Wahlbeteiligung, nach Gemeinden/Bezirk/Canton* (http://www.bfs.admin.ch/bfs/portal/de/index/themen/17/02/blank/data/04/03.html) (German) accessed 28 May 2010

[17] Elezioni cantonali: Gran Consiglio, Consiglio di Stato (http://www.ti.ch/DFE/USTAT/DATI_RIEPILOGHI/default_x.asp?menu=17&key1_testo=Politica&key1=17&key2=02&key3=02) (Italian) accessed 23 November 2010

[18] Swiss Federal Statistical Office - Statweb (http://www.media-stat.admin.ch/stat/pendler/pop.php) (German) accessed 24 June 2010

[19] Settori alberghiero e paralberghiero (http://www.ti.ch/DFE/USTAT/DATI_RIEPILOGHI/default_x.asp?menu=10&key1_testo=Turismo&key1=10&key2=03&key3=01) (Italian) accessed 23 November 2010

[20] EDK/CDIP/IDES (2010). KANTONALE SCHULSTRUKTUREN IN DER SCHWEIZ UND IM FÜRSTENTUM LIECHTENSTEIN / STRUCTURES SCOLAIRES CANTONALES EN SUISSE ET DANS LA PRINCIPAUTÉ DU LIECHTENSTEIN (http://edudoc.ch/record/35128/files/Schulsystem_alle.pdf) (Report). . Retrieved 24 June 2010.

[21] Allievi e studenti, secondo il genere di scuola, anno scolastico 2009/2010 (http://www.ti.ch/DFE/USTAT/DATI_RIEPILOGHI/default_x.asp?menu=15&key1_testo=Formazione e scienza&key1=15&key2=02&key3=01) (Italian) accessed 23 November 2010

External links

- Official website (http://www.camorino.ch) (Italian)
- *Camorino* in German (http://www.hls-dhs-dss.ch/textes/d/D2033.php), French (http://www.hls-dhs-dss.ch/textes/f/F2033.php) and Italian (http://www.hls-dhs-dss.ch/textes/i/I2033.php) in the online *Historical Dictionary of Switzerland*.

Cantons_of_Switzerland

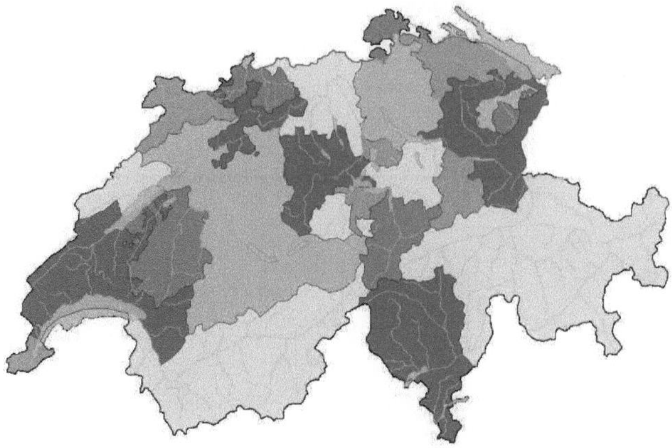

Valais

Ticino

Graubünden (Grisons)

Geneva

Vaud

Neuchâtel

Jura

Bern

Thurgau

Zurich

Aargau

Lucerne

Solothurn

Basel-Land

Schaffhausen

Uri

Schwyz

Glarus

St. Gallen

AI

AR

Obwalden

Nidwalden

Zug

Fribourg

Basel-Stadt

France

Italy

Liecht.

Austria

Germany

The 26 **cantons of Switzerland** are the member states of the federal state of Switzerland. Each canton was a fully sovereign state[1] with its own border controls, army and currency from the Treaty of Westphalia (1648) until the establishment of the Swiss federal state in 1848. The most recently created canton is the Canton of Jura, which separated from the Canton of Bern in 1979.[2]

The name is derived from the French language word *canton* meaning *corner* or *district* (from which the term Cantonment is also derived).

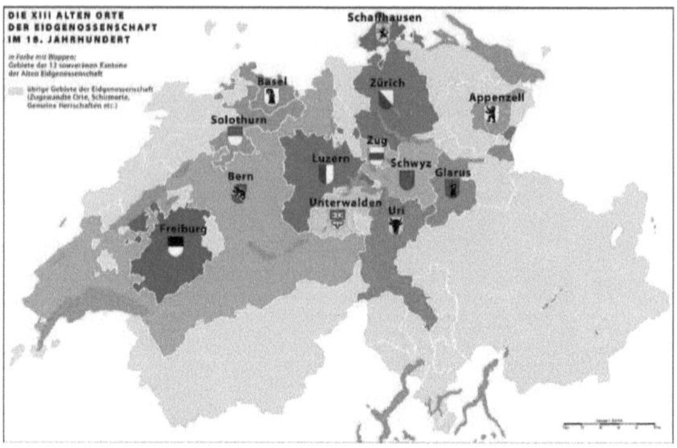

The 13 cantons of the Old Swiss Confederacy (1513-1798).

History

In the 16th century, the Old Swiss Confederacy was composed of 13 sovereign cantons, and there were two different kinds: six land (or forest) cantons and seven city (or urban) cantons. Though they were technically part of the Holy Roman Empire, they had become *de facto* independent when the Swiss defeated Emperor Maximillian in 1499.[] The six forest cantons were democratic republics, whereas the seven urban cantons were oligarchic republics controlled by noble families.

Constitution

Each canton has its own constitution, legislature, government and courts.[3] Most of the cantons' legislatures are unicameral parliaments, their size varying between 58 and 200 seats. A few legislatures are general assemblies known as *Landsgemeinden*. The cantonal governments consist of either five or seven members, depending on the canton.[4] For the names of the institutions, see List of legislative and executive councils of the Cantons of Switzerland.

The Swiss Federal Constitution declares the cantons to be sovereign to the extent their sovereignty is not limited by federal law.[3] The cantons also retain all powers and competencies not delegated to the Confederation by the Constitution. Most significantly, the cantons are responsible for healthcare, welfare, law enforcement and public education; they also retain the power of taxation. The cantonal constitutions determine the degree of autonomy accorded to the municipalities, which varies but almost always includes the power to levy taxes and pass municipal laws. The sizes of the cantons vary from 37 km² to 7,105 km²; the populations vary from 15,471 to 1,244,400.

Direct democracy

As on the federal level, all cantons provide for (half-) direct democracy. Citizens may demand a popular vote to amend the cantonal constitution or laws, or to veto laws or spending bills passed by the parliament. General popular assemblies (*Landsgemeinde*) are now limited to the cantons of Appenzell Innerrhoden and Glarus. In all other cantons democratic rights are exercised by secret ballot.

List

The cantons are listed in the order given in the federal constitution.[a]

Coat of arms	Abbr	Canton	Since	Capital	Population[1]	Area (km²) [2]	Density (per km²) [3]	No. munic.[4]	Official languages
	ZH	**Zurich**	1351	Zurich	1406083		701	171	German
	BE	**Bern**	1353	Bern	985046		158	383	German, French
	LU	**Lucerne**	1332	Lucerne	381966		233	87	German
	UR	**Uri**	1291[5]	Altdorf	35382		33	20	German
	SZ	**Schwyz**	1291[5]	Schwyz	147904		143	30	German
	OW	**Obwalden**	1291[5] or 1315 (as part of Unterwalden)	Sarnen	35878		66	7	German
	NW	**Nidwalden**	1291[5] (as Unterwalden)	Stans	41311		138	11	German
	GL	**Glarus**	1352	Glarus	39217		51	3	German
	ZG	**Zug**	1352	Zug	116559		416	11	German
	FR	**Fribourg**	1481	Fribourg	284668		141	167	French, German
	SO	**Solothurn**	1481	Solothurn	259836		308	122	German
	BS	**Basel-Stadt**	1501 (as Basel until 1833/1999)	Basel	194090		5,072	3	German
	BL	**Basel-Landschaft**	1501/1833[6]	Liestal	277973		502	86	German
	SH	**Schaffhausen**	1501	Schaffhausen	77139		246	27	German
	AR	**Appenzell Ausserrhoden**	1513 [7]	Herisau[5]	53313		220	20	German
	AI	**Appenzell Innerrhoden**	1513[7]	Appenzell	15789		87	6	German
	SG	**St. Gallen**	1803[8]	St. Gallen	483101		222	85	German
	GR	**Graubünden**	1803[9]	Chur	193388		26	180	German, Romansh, Italian
	AG	**Aargau**	1803	Aarau	627893		388	220	German

	TG	**Thurgau**	1803[10]	Frauenfeld[6]	254528		229	80	German
	TI	**Ticino**	1803[11]	Bellinzona	336943		110	157	Italian
	VD	**Vaud**	1803[12]	Lausanne	725944		188	339	French
	VS	**Valais**	1815[13]	Sion	317022		53	143	French, German
	NE	**Neuchâtel**	1815/1857[14]	Neuchâtel	173183		206	53	French
	GE	**Geneva**	1815	Geneva	472530		1,442	45	French
	JU	**Jura**	1979[15]	Delémont	70542		82	64	French
	CH	**Switzerland**		Bern	7,968,705		174	2,596	German, French, Italian, Romansh

The two-letter abbreviations for Swiss cantons are widely used, e.g. on car license plates. They are also used in the ISO 3166-2 codes of Switzerland with the prefix "CH-" (*Confœderatio Helvetica*, Switzerland), e.g. CH-SZ for the canton of Schwyz.

Half-cantons

Six of the 26 cantons are traditionally, but no longer officially, called "half-cantons" (German: *Halbkanton*, French: *demi-canton*, Italian: *semicantone*), reflecting a history of mutual association or partition.

The half-cantons are identified in the first article of the Swiss Federal Constitution of 1999 by being joined to their other "half" with the conjunction "and":

> The People and the Cantons of Zurich, Bern, Lucerne, Uri, Schwyz, Obwalden and Nidwalden, Glarus, Zug, Fribourg, Solothurn, Basel-Stadt and Basel-Landschaft, Schaffhausen, Appenzell Ausserrhoden and Appenzell Innerrhoden, St. Gallen, Graubünden, Aargau, Thurgau, Ticino, Vaud, Valais, Neuchâtel, Geneva, and Jura form the Swiss Confederation.
>
> —Article 1 of the Federal Constitution of the Swiss Confederation (underlining not in original)[16]

The 1999 constitutional revision retained this distinction, on the request of the six cantonal governments, as a way to mark the historic association of the half-cantons to each other.[17] In contrast, the first article of the 1848 and 1874 constitutions constituted the Confederation as the union of "twenty-two sovereign cantons",[18] referring to the half-cantons as "Unterwalden (above and beneath the woods)", "Basel (city and country)" and "Appenzell (both Rhoden)".[19] While the older constitutions referred to these states as "half-cantons", a term that remains in popular use, the 1999 revision and official terminology since then use the appellation "cantons with half a cantonal vote".[20]

With their mutual association a purely historical matter, the half-cantons are since 1848 equal to the other cantons in all but two respects:[21]

- They elect only one member of the Council of States instead of two (Cst. art. 150 par. 2).
- In popular referendums about constitutional amendments, which require for adoption a national popular majority as well as the assent of a majority of the cantons (*Ständemehr / majorité des cantons*), the result of the half-cantons' popular vote counts only one half of that of the other cantons (Cst. arts. 140, 142). This means that for purposes of a constitutional referendum, at least 12 out of a total of 23 cantonal popular votes must support the amendment.[22]

The reasons for the association between the three pairs of half-cantons are varied:

Caricature of the division of Basel, 1833

- *Unterwalden* never consisted of a single unified jurisdiction. Originally, Obwalden, Nidwalden, and the Abbey of Engelberg formed distinct communities. The collective term *Unterwalden* remains in use, however, for the area that partook in the creation of the original Swiss confederation in 1291 with Uri and Schwyz. The Federal Charter of 1291 called for representatives from each of the three "areas".[23] [24]
- The canton of Appenzell divided itself into an "inner" and "outer" half ("Rhoden") as a consequence of the Reformation in Switzerland in 1597:[25] Appenzell Innerrhoden (Catholic) and Appenzell Ausserrhoden (Protestant).
- The canton of Basel divided itself as a consequence of a revolt of the Basel countryside in 1833, in order to promote equality among its citizenry, combating claims between rural and city residents over preferential status:[26] Basel-Landschaft and Basel-Stadt.

Names in national languages

(Names appear in bold when corresponding to the cantonal official language)

Abbr	Common English	Other English forms	German	French	Italian	Romansh
AG	Aargau	Argovia	**Aargau**	Argovie	Argovia	Argovia
AI	Appenzell Innerrhoden	Appenzell Inner-Rhodes	**Appenzell Innerrhoden**	Appenzell Rhodes-Intérieures	Appenzello Interno	Appenzell dadens
AR	Appenzell Ausserrhoden	Appenzell Outer-Rhodes	**Appenzell Ausserrhoden**	Appenzell Rhodes-Extérieures	Appenzello Esterno	Appenzell dador
BS	Basel-Stadt	Basle-City	**Basel-Stadt**	Bâle-Ville	Basilea-Città	Basilea-Citad
BL	Basel-Landschaft	Basle-Country	**Basel-Landschaft**	Bâle-Campagne	Basilea-Campagna	Basilea-Champagna
BE	Bern	Berne	**Bern**	**Berne**	Berna	Berna
FR	Fribourg	Friburg	**Freiburg**	**Fribourg**	Friborgo	Friburg
GE	Geneva	-	Genf	**Genève**	Ginevra	Genevra
GL	Glarus	Glaris	**Glarus**	Glaris	Glarona	Glaruna
GR	Graubünden	Grisons	**Graubünden**	Grisons	**Grigioni**	**Grischun**
JU	Jura	-	Jura	**Jura**	Giura	Giura
LU	Lucerne	-	**Luzern**	Lucerne	Lucerna	Lucerna
NE	Neuchâtel	-	Neuenburg	**Neuchâtel**	Neuchâtel	Neuchâtel
NW	Nidwalden	Nidwald	**Nidwalden**	Nidwald	Nidvaldo	Sutsilvania
OW	Obwalden	Obwald	**Obwalden**	Obwald	Obvaldo	Sursilvania
SH	Schaffhausen	Schaffhouse	**Schaffhausen**	Schaffhouse	Sciaffusa	Schaffusa
SZ	Schwyz	-	**Schwyz**	Schwyz (or Schwytz)	Svitto	Sviz
SO	Solothurn	Soleure	**Solothurn**	Soleure	Soletta	Soloturn
SG	St. Gallen	St. Gall	**St. Gallen**	Saint-Gall	San Gallo	Son Gagl

TG	Thurgau	Thurgovia	**Thurgau**	Thurgovie	Turgovia	Turgovia
TI	Ticino	Tessin	Tessin	Tessin	**Ticino**	Tessin
UR	Uri	-	**Uri**	Uri	Uri	Uri
VS	Valais	Wallis	**Wallis**	**Valais**	Vallese	Vallais
VD	Vaud	-	Waadt	**Vaud**	Vaud	Vad
ZG	Zug	-	**Zug**	Zoug	Zugo	Zug
ZH	Zurich	-	**Zürich**	Zurich	Zurigo	Turitg

Admission of new cantons

The enlargement of Switzerland by way of the admission of new cantons ended in 1815. After a failed attempt of Vorarlberg to join Switzerland in 1919, the idea of resuming Swiss enlargement was revived in 2010 by a parliamentary motion that would allow the accession of regions bordering on Switzerland.

See also

- List of legislative and executive councils of the Cantons of Switzerland
- Data codes for Switzerland#Cantons
- List of cantons of Switzerland by elevation
- Flags of Swiss cantons
- Cantonal bank, a commercial bank (at least partially) owned by the canton

Notes

1. This is the order generally used in Swiss official documents. At the head of the list are the three city cantons that were considered preeminent in the Old Swiss Confederacy; the other cantons are listed in order of accession to the Confederation. This traditional order of precedence among the cantons has no practical relevance in the modern federal state, in which the cantons are equal to one another, although it still determines formal precedence among the cantons' officials (see Swiss order of precedence).
2. as of 5 April 2009
3. km²
4. Per km², based on 2000 population
5. As of 31 December 2007,
6. Seat of government and parliament is Herisau, the seat of the judicial authorities is Trogen
7. Seat of parliament half-yearly alternates between Frauenfeld and Weinfelden

References

- (German). Cited as *Ehrenzeller.*
- Cited as *Häfelin.*

[1] *Cantons, In the Old Confederation until 1798* in German (http://www.hls-dhs-dss.ch/textes/d/D26414-1-1.php), French (http://www.hls-dhs-dss.ch/textes/f/F26414-1-1.php) and Italian (http://www.hls-dhs-dss.ch/textes/i/I26414-1-1.php) in the online *Historical Dictionary of Switzerland.*

[2] *Jura (Canton)* in German (http://www.hls-dhs-dss.ch/textes/d/D7399.php), French (http://www.hls-dhs-dss.ch/textes/f/F7399.php) and Italian (http://www.hls-dhs-dss.ch/textes/i/I7399.php) in the online *Historical Dictionary of Switzerland.*

[3] *Cantons, In the Federal State since 1848* in German (http://www.hls-dhs-dss.ch/textes/d/D26414-1-4.php), French (http://www.hls-dhs-dss.ch/textes/f/F26414-1-4.php) and Italian (http://www.hls-dhs-dss.ch/textes/i/I26414-1-4.php) in the online *Historical Dictionary of Switzerland.*

[4] Swiss Government website (http://www.ch.ch/karte/index.html?lang=en) with links to each cantonal government, accessed 11 November 2008
[5] founding forest-canton, foundation date traditionally given as either 1307, 1304 or 1291 (see Foundation of the Old Swiss Confederacy).
[6] part of Basel until 1833/1999
[7] part of Appenzell until 1597/1999
[8] Act of Mediation, formed out of the Canton of Säntis and the northern half of the Canton of Linth.
[9] Act of Mediation; formerly the Canton of Raetia, comprising the earlier Three Leagues.
[10] coterminous with the canton of Thurgau of the Helvetic Republic (1798), formerly a condominium.
[11] combining the former cantons of Bellinzona and Lugano; see *Ennetbirgische Vogteien.*
[12] Act of Mediation, formerly Canton of Léman.
[13] Restoration, formerly the Simplon *département*
[14] claimed by Frederick William III of Prussia until the Neuchâtel Crisis of 1856–1857.
[15] seceded from Berne
[16] Federal Constitution of the Swiss Confederation of 18 April 1999, SR/RS 101 (E (http://www.admin.ch/ch/e/rs/c101.html)· D (http://www.admin.ch/ch/d/sr/c101.html)· F (http://www.admin.ch/ch/f/rs/c101.html)· I (http://www.admin.ch/ch/i/rs/c101.html)), art. 1 (E (http://www.admin.ch/ch/e/rs/101/a1.html)· D (http://www.admin.ch/ch/d/sr/101/a1.html)· F (http://www.admin.ch/ch/f/rs/101/a1.html)· I (http://www.admin.ch/ch/i/rs/101/a1.html))
[17] Felix Hafner / Rainer J. Schweizer in *Ehrenzeller*, Art. 1 N 2; Häfelin, N 966.
[18] Twenty-three after the creation of the Canton of Jura in 1978.
[19] Bundesverfassung der Schweizerischen Eidgenossenschaft vom 29. Mai 1874 (http://www.verfassungen.de/ch/verf74-i.htm), Bundesverfassung der Schweizerischen Eidgenossenschaft vom 12. September 1848 (http://www.verfassungen.de/ch/verf48.htm) (German); author's translation.
[20] Felix Hafner / Rainer J. Schweizer in *Ehrenzeller*, Art. 1 N 10; Häfelin, N 963
[21] Häfelin, N 963, 967
[22] Häfelin, N 950
[23] Pacte fédéral du 1er (http://www.admin.ch/org/polit/00056/index.html?lang=fr) août 1291] sur Admin.ch "vallée inférieure d'Unterwald" signifie Nidwald.
[24] Pacte fédéral du 1er août 1291 (http://hypo.ge.ch/www/cliotexte//html/suisse.histoire.1291.html) sur Cliotexte
[25] Réforme catholique, Contre-Réforme et scission (http://hls-dhs-dss.ch/textes/f/F7389-1-10.php) Article du dictionnaire historique de la Suisse
[26] De la République helvétique à la division du canton (1798-1833) (http://hls-dhs-dss.ch/textes/f/F7387-2-2.php) Article du dictionnaire historique de la Suisse

External links

- Swissworld.org (http://www.swissworld.org/en/politics/general_information/the_cantons/) – The cantons of Switzerland
- GeoPuzzle (http://www.geomit.ch/GeoPuzzles/SchweizPuzzle/schweiz.html) – Assemble cantons on a Swiss map
- Badac (http://www.badac.ch) – Database on Swiss cantons and cities (French/German)

Bellinzona_(district)

The district of **Bellinzona** (Italian: *Distretto di Bellinzona* also called **Belinzonese**) is a district of Canton Ticino, Switzerland. It has a population of 48871 (as of 31 December 2011).[1]

Geography

The district has an area, as of 1997, of 211.88 square kilometers (sq mi). Of this area, 27.98 km^2 (sq mi) or 13.2% is used for agricultural purposes, while 145.46 km^2 (sq mi) or 68.7% is forested. Of the rest of the land, 17.51 km^2 (sq mi) or 8.3% is settled (buildings or roads), 3.27 km^2 (sq mi) or 1.5% is either rivers or lakes and 15.4 km^2 (sq mi) or 7.3% is unproductive land.

Of the built up area, housing and buildings made up 4.0% and transportation infrastructure made up 2.5%. Out of the forested land, 59.7% of the total land area is heavily forested and 2.2% is covered with orchards or small clusters of trees. Of the agricultural land, 7.5% is used for growing crops, while 1.7% is used for orchards or vine crops and 4.0% is used for alpine pastures. All the water in the district is flowing water. Of the unproductive areas, 5.2% is unproductive vegetation and 2.1% is too rocky for vegetation.[2]

Demographics

Of the Swiss national languages (as of 2000), 1,843 speak German, 492 people speak French, 38,109 people speak Italian, and 42 people speak Romansh. The remainder (2,912 people) speak another language.[3]

As of 2008, the gender distribution of the population was 48.3% male and 51.7% female. The population was made up of 16,770 Swiss men (34.6% of the population), and 6,606 (13.6%) non-Swiss men. There were 19,192 Swiss women (39.6%), and 5,838 (12.1%) non-Swiss women.[4]

In 2008 there were 364 live births to Swiss citizens and 98 births to non-Swiss citizens, and in same time span there were 329 deaths of Swiss citizens and 51 non-Swiss citizen deaths. Ignoring immigration and emigration, the population of Swiss citizens increased by 35 while the foreign population increased by 47. There were 6 Swiss men and 4 Swiss women who emigrated from Switzerland. At the same time, there were 188 non-Swiss men and 157 non-Swiss women who immigrated from another country to Switzerland. The total Swiss population change in 2008 (from all sources) was an increase of 749 and the non-Swiss population change was a decrease of 91 people. This represents a population growth rate of 1.4%.[5]

The age distribution, as of 2009, in the district is; 4,644 children or 9.6% of the population are between 0 and 9 years old and 4,796 teenagers or 9.9% are between 10 and 19. Of the adult population, 5,627 people or 11.6% of the population are between 20 and 29 years old. 7,097 people or 14.7% are between 30 and 39, 8,137 people or 16.8% are between 40 and 49, and 6,291 people or 13.0% are between 50 and 59. The senior population distribution is 5,541 people or 11.4% of the population are between 60 and 69 years old, 3,732 people or 7.7% are between 70 and 79, there are 2,541 people or 5.2% who are over 80.[4]

In 2000 there were 15,519 single family homes (or 45.9% of the total) out of a total of 33,816 inhabited buildings. There were 4,707 two family buildings (13.9%) and 10,816 multi-family buildings (32.0%). There were also 2,774 buildings in the district that were multipurpose buildings (used for both housing and commercial or another purpose).[6] In 2000 there were 22,466 apartments in the district. The most common apartment size was the 4 room apartment of which there were 7,594. There were 1,109 single room apartments and 4,040 apartments with five or more rooms.[7] Of these apartments, a total of 18,452 apartments (82.1% of the total) were permanently occupied, while 3,421 apartments (15.2%) were seasonally occupied and 593 apartments (2.6%) were empty.[7]

The historical population is given in the following table:[8]

year	population
1850	11,582
1880	13,597
1900	17,742
1950	24,083
1960	26,681
1980	37,956
1990	40,618
2000	43,398

Politics

In the 2007 federal election the most popular party was the FDP which received 32.78% of the vote. The next three most popular parties were the CVP (22.94%), the SP (20.19%) and the Ticino League (11.12%). In the federal election, a total of 13,901 votes were cast, and the voter turnout was 48.7%.[9]

In the 2007 *Ticino Gran Consiglio* election, there were a total of 28,519 registered voters in the district, of which 19,508 or 68.4% voted. 310 blank ballots and 37 null ballots were cast, leaving 19,161 valid ballots in the election. The most popular party was the PLRT which received 5,371 or 28.0% of the vote. The next three most popular parties were; the SSI (with 3,541 or 18.5%), the PPD+GenGiova (with 3,405 or 17.8%) and the PS (with 3,193 or 16.7%).[10]

In the 2007 *Ticino Consiglio di Stato* election, 209 blank ballots and 56 null ballots were cast, leaving 19,247 valid ballots in the election. The most popular party was the PLRT which received 4,975 or 25.8% of the vote. The next three most popular parties were; the PS (with 3,799 or 19.7%), the PPD (with 3,459 or 18.0%) and the SSI (with 3,162 or 16.4%).[10]

Religion

From the 2000 census, 33,661 or 77.6% were Roman Catholic, while 1,908 or 4.4% belonged to the Swiss Reformed Church. There are 5,346 individuals (or about 12.32% of the population) who belong to another church (not listed on the census), and 2,483 individuals (or about 5.72% of the population) did not answer the question.[3]

Education

The district had a total of 7,949 students (as of 2009). The Ticino education system provides up to three years of non-mandatory kindergarten and there were 1,296 children in kindergarten. The primary school program lasts for five years and includes both a standard school and a special school. In the district, 2,357 students attended the standard primary schools and 153 students attended the special school. In the lower secondary school system, students either attend a two-year middle school followed by a two-year pre-apprenticeship or they attend a four-year program to prepare for higher education. There were 1,911 students in the two-year middle school and 11 in their pre-apprenticeship, while 749 students were in the four-year advanced program.

The upper secondary school includes several options, but at the end of the upper secondary program, a student will be prepared to enter a trade or to continue on to a university or college. In Ticino, vocational students may either attend school while working on their internship or apprenticeship (which takes three or four years) or may attend school followed by an internship or apprenticeship (which takes one year as a full-time student or one and a half to two years as a part-time student).[11] There were 438 vocational students who were attending school full-time and

905 who attend part-time.

The professional program lasts three years and prepares a student for a job in engineering, nursing, computer science, business, tourism and similar fields. There were 129 students in the professional program.[12]

Circles and municipalities

Circolo di Bellinzona			
Coat of arms	**Municipality**	**Population (31 December 2011)**[1]	**Area km²**[13]
	Bellinzona	17544	19.15
	Arbedo-Castione	4380	21.28
	Lumino	1362	9.95
	Total	23286	**50.38**

Circolo del Ticino			
Coat of arms	**Municipality**	**Population (31 December 2011)**[1]	**Area km²**[13]
	Monte Carasso	2725	9.64
	Gudo	770	9.94
	Sementina	3111	8.25
	Gorduno	677	9.22
	Gnosca	705	7.45
	Preonzo	588	16.44
	Moleno	106	7.49
	Total	8682	**68.43**

Circolo di Giubiasco			
Coat of arms	**Municipality**	**Population (31 December 2011)**[1]	**Area km²**[13]
	Giubiasco	8364	6.23
	Pianezzo	579	8.02
	Sant'Antonio	214	33.58
	Camorino	2697	8.28
	Sant'Antonino	2264	6.59
	Cadenazzo	2401	8.37
	Isone	384	12.82
	Total	16903	**83.89**

References

[1] Swiss Federal Statistics Office – STAT-TAB (http://www.pxweb.bfs.admin.ch/Dialog/varval.asp?ma=px-d-01-2A01&ti=Ständige+und+Nichtständige+Wohnbevölkerung+nach+Region,+Geschlecht,+Nationalität+und++Alter&path=../Database/German_01 - Bevölkerung/01.2 - Bevölkerungsstand und -bewegung/&lang=1&prod=01&openChild=true&secprod=2) *Ständige und Nichtständige Wohnbevölkerung nach Region, Geschlecht, Nationalität und Alter* (German) accessed 3 October 2012

[2] Altitudine, superficie, secondo il genere di utilizzazione, rilevazione 1992/1997, e densità della popolazione, nel 2000 (http://www.ti.ch/DFE/USTAT/DATI_RIEPILOGHI/default_x.asp?menu=02&key1_testo=Territorio e ambiente&key1=02&key2=01&key3=01) (Italian) accessed 25 October 2010

[3] Popolazione residente, secondo la lingua principale e la religione, nel 2000 (http://www.ti.ch/DFE/USTAT/DATI_RIEPILOGHI/default_x.asp?menu=01&key1_testo=Popolazione&key1=01&key2=05&key3=00) (Italian) accessed 23 November 2010

[4] 01.02.03 Popolazione residente permanente (http://www.ti.ch/DFE/USTAT/DATI_RIEPILOGHI/default_x.asp?menu=01&key1_testo=Popolazione&key1=01&key2=02&key3=03) (Italian) accessed 23 November 2010

[5] Swiss Federal Statistical Office - Superweb database - Gemeinde Statistics 1981-2008 (http://www.bfs.admin.ch/bfs/portal/de/index/infothek/onlinedb/superweb/login.html) (German) accessed 19 June 2010

[6] 09.02.01 Edifici (http://www.ti.ch/DFE/USTAT/DATI_RIEPILOGHI/default_x.asp?menu=09&key1_testo=Costruzioni e abitazioni&key1=09&key2=02&key3=01) (Italian) accessed 23 November 2010

[7] 09.02.02 Abitazioni (http://www.ti.ch/DFE/USTAT/DATI_RIEPILOGHI/default_x.asp?menu=09&key1_testo=Costruzioni e abitazioni&key1=09&key2=02&key3=02) (Italian) accessed 23 November 2010

[8] Popolazione residente, dal 1850 (http://www.ti.ch/DFE/USTAT/DATI_RIEPILOGHI/default_x.asp?menu=01&key1_testo=Popolazione&key1=01&key2=02&key3=02) (Italian) accessed 23 November 2010

[9] Swiss Federal Statistical Office, *Nationalratswahlen 2007: Stärke der Parteien und Wahlbeteiligung, nach Gemeinden/Bezirk/Canton* (http://www.bfs.admin.ch/bfs/portal/de/index/themen/17/02/blank/data/04/03.html) (German) accessed 28 May 2010

[10] Elezioni cantonali: Gran Consiglio, Consiglio di Stato (http://www.ti.ch/DFE/USTAT/DATI_RIEPILOGHI/default_x.asp?menu=17&key1_testo=Politica&key1=17&key2=02&key3=02) (Italian) accessed 23 November 2010

[11] EDK/CDIP/IDES (2010). KANTONALE SCHULSTRUKTUREN IN DER SCHWEIZ UND IM FÜRSTENTUM LIECHTENSTEIN / STRUCTURES SCOLAIRES CANTONALES EN SUISSE ET DANS LA PRINCIPAUTÉ DU LIECHTENSTEIN (http://edudoc.ch/record/35128/files/Schulsystem_alle.pdf) (Report).. Retrieved 24 June 2010.

[12] Allievi e studenti, secondo il genere di scuola, anno scolastico 2009/2010 (http://www.ti.ch/DFE/USTAT/DATI_RIEPILOGHI/default_x.asp?menu=15&key1_testo=Formazione e scienza&key1=15&key2=02&key3=01) (Italian) accessed 23 November 2010
[13] Swiss Federal Statistical Office-Land Use Statistics (http://www.bfs.admin.ch/bfs/portal/de/index/themen/02/03/blank/data/gemeindedaten.html) 2009 data (German) accessed 25 March 2010

- Official list of municipalities, circles and districts of Ticino (http://www.ti.ch/CAN/argomenti/Legislaz/rleggi/rl/dati_rl/f/f2.htm)

Communes_of_Switzerland

Communes (German: *Gemeinden* / French: *communes* / Italian: *comuni* / Romansh: *vischnancas*), also known as municipalities, are the smallest government division in Switzerland, numbering 2,596 (as of February 2010).[1] While many have a population of a few hundred citizens, the largest cities such as Zürich or Geneva also have the legal status of municipalities. The area of the municipalities varies between 0.28 km² (Ponte Tresa, Ticino) and 430 km² (Glarus Süd, Glarus).

Map of Switzerland showing municipal boundaries (2012)

Each canton defines the responsibilities of its constituent communes. These may include providing local government services such as education, medical and social services, public transportation, and tax collection. The degree of centralization varies from one canton to another.

Communes are generally governed by a council (sometimes called Municipality) headed by a mayor as executive and the town meeting as legislature. Most cantons leave the option to larger municipalities to opt for a city parliament. In some cantons, foreign persons that have dwelled for a certain time in Switzerland are also allowed to participate in the municipal politics.

Swiss citizenship is based on the citizenship of a municipality. Every Swiss is citizen of one or several municipalities (known as the place of origin, *lieu d'origine*, or *Heimatort*). As on the cantonal and federal level, citizens enjoy their political rights, including the direct democratic ones, in their municipality.

Communes are financed through direct taxes (such as the income tax), with rates varying more or less within a framework set by the canton. As among the cantons, there is a tax transfer among the communes to balance various levels of tax income.

Many municipalities are having difficulties maintaining the civil services they need to perform the duties they are required to do. In an effort to reduce expenses, some municipalities are combining together (through mergers or the creation of special-purpose districts). This restructuring is generally encouraged by the cantonal governments and the rate of these unions is increasing.

"Cities" (*villes* or *Städte*) are the municipalities with more than 10'000 inhabitants, or smaller places which had medieval town rights. There is no specific designation for smaller communities such as "village" or "town".

Population	Number of municipalities in 2004 (%)[2]
>20,000	30 (1.1%)
10,000–19,999	89 (3.2%)
5000–9999	180 (6.6%)
1000–4999	1025 (37.4%)
500–999	555 (20.3%)
<500	861 (31.4%)
Total	2740 (100%)

Communes are numbered by the Swiss Federal Office for Statistics (see *Community Identification Number#Switzerland*). One or more postal codes (PLZ/NPA) can by assigned to a municipality or shared with other municipalities.

See also: List of cities in Switzerland

Lists of communes by canton

- Aargau: *Einwohnergemeinden*
- Appenzell I.R.: *Bezirke*
- Appenzell A.R.: *Einwohnergemeinden*
- Basel-City: *Einwohnergemeinden*
- Basel-Country: *Einwohnergemeinden*
- Bern: *Einwohnergemeinden* or *communes municipales*
- Fribourg: *communes* or *Gemeinden*
- Geneva: *communes*
- Glarus: *Ortsgemeinden*
- Grisons: *politische Gemeinden, Vischnancas politicas* or *comuni politici*
- Jura: *communes*
- Lucerne: *Einwohnergemeinden*
- Neuchâtel: *communes*
- Nidwalden: *Gemeinden*
- Obwalden: *Einwohnergemeinden*
- Schaffhausen: *Einwohnergemeinden*
- Schwyz: *Gemeinden*
- Solothurn: *Einwohnergemeinden*
- St. Gallen: *Politische Gemeinden*
- Thurgau: *politische Gemeinden*
- Ticino: *comuni politici*
- Uri: *Einwohnergemeinde*
- Valais: *communes municipales* or *Einwohnergemeinden*
- Vaud: *communes*
- Zug: *Einwohnergemeinde*
- Zürich: *Politische Gemeinden*

See cantons of Switzerland for the number of municipalities per canton.

List of communes by population

Largest municipalities (2011)[]

Rank	City	Canton	Pop.
1	Zürich	Zürich	376,990
2	Geneva	Geneva	188,234
3	Basel	Basel-City	164,516
4	Lausanne	Vaud	129,383
5	Bern	Bern	125,681
6	Winterthur	Zürich	103075
7	Lucerne	Lucerne	78,093
8	St. Gallen	St. Gallen	73,505
9	Lugano	Ticino	55,151

10	Biel/Bienne	Bern	51635

Smallest municipalities (2011)[]

Rank	City	Canton	Pop.
1	Corippo	Ticino	12
2	Martisberg	Valais	19
3	Mulegns	Graubünden	29
4	St. Martin	Graubünden	31
5	Selma	Graubünden	33
6	Pigniu	Graubünden	33
6	Bister	Valais	33
8	Gresso	Ticino	34
9	Cauco	Graubünden	35
10	Monible	Bern	37

History

The beginnings of the modern municipality system date back to the Helvetic Republic. Under the Old Swiss Confederacy, citizenship was granted by each town and village to only residents. These citizens enjoyed access to community property and in some cases additional protection under the law. Additionally, the urban towns and the rural villages had differing rights and laws. The creation of a uniform Swiss citizenship, which applied equally for citizens of the old towns and their tenants and servants, led to conflict. The wealthier villagers and urban citizens held rights rights to forests, common land and other municipal property which they did not want to share with the "new citizens", who were generally poor. The compromise solution, which was written into the municipal laws of the Helvetic Republic, is still valid today. Two politically separate but often geographically similar organizations were created. The first, the so-called municipality, was a political community formed by election and its voting body consists of all resident citizens. However, the community land and property remained with the former local citizens who were gathered together into the *Bürgergemeinde*. During the Mediation era (1803–1814), and especially during the Restoration era (1814–1830), many of the gains toward uniform citizenship were lost. Many political municipalities were abolished and limits were placed on the exercise of political rights for everyone except the members of the *Bürgergemeinde*. In the Regeneration era (1830–1848), the liberal revolutions of the common people helped to restore some rights again in a few cantons. In other cantons, the *Bürgergemeinden* were able to maintain power as political communities. In the city of Zurich it wasn't until the Municipal Act of 1866 that the political municipality came back into existence.[3]

The relationship between the political municipality and the *Bürgergemeinde* was often dominated by the latter's ownership of community property. Often the administration and profit from the property were totally held by the *Bürgergemeinden*, leaving the political municipality dependent on the *Bürgergemeinde* for money and use of the property. It wasn't until the political municipality acquired rights over property that served the public (such as schools, fire stations, etc.) and taxes, that they obtained full independence. For example, in the city of Bern, it wasn't until after the property division of 1852 that the political municipality had the right to levy taxes.[3]

It wasn't until the Federal Constitution of 1874 that all Swiss citizens were granted equal political rights on local and Federal levels. This revised constitution finally removed all the political voting and electoral body rights from the *Bürgergemeinde*. In the cities, the percentage of members in the *Bürgergemeinde* in the population was reduced as a result of increasing emigration to the cities. This led to the *Bürgergemeinde* losing its former importance to a large extent. However, the *Bürgergemeinde* has remained, and it includes all individuals who are citizens of the *Bürgergemeinde*, usually by having inherited the *Bürgerrecht* (citizenship), regardless of where they were born or where they may currently live. Instead of the place of birth, Swiss legal documents, e.g. passports, contain the *Bürgerort* (place of citizenship). The *Bürgergemeinde* also often holds and administers the common property in the village for the members of the community.[3]

Other types of communes

In addition to the political communes or municipalities, a number of other communes exist in Switzerland. These include:

- Bürgergemeinde (also: Burgergemeinde, Ortsgemeinde, Ortsbürgergemeinde, Tagwen, bourgeoisie, commune bourgeoise, vischnanca burgaisa), a statutory corporation that includes everyone who is a citizen of a commune and has the *Heimatrecht* (home right) in that commune regardless of where they may currently live. Until the 19th Century this *Heimatrecht* included rights to use the commons, which were administered by the Bürgergemeinde. Modernly, some Bürgergemeinden may still control common property, but the *Heimatrecht* and associated *Heimatort* is used just as place of birth in other countries.
- Gemischte Gemeinde (mixed communes), found in the Canton of Jura and portions of the Canton of Bern, a combination of a Bürgergemeinde and a political commune.
- Korporationsgemeinde, a legally recognized cooperative in Central Switzerland that controls some land and is responsible for its members support.
- Kirchgemeinde, a parish for members of a large church (generally Roman Catholic or Swiss Reformed. There may be two or more *Kirchgemeinden* in a single municipality.
- Schulgemeinde, similar to a School district.
- Bäuert, in the Berner Oberland or Graubünden) a small farming community. It is a type of agricultural cooperative with shared equipment and land.[4]
- Degagna, in the Leventina valley in the Canton of Ticino. It manages shared pastures, fields and woods as well as maintaining roads that cross the common land.[5]

See also

- List of municipalities of Switzerland
- List of twin towns and sister cities in Switzerland
- Former municipalities of Switzerland

Notes and references

[2] Official list of Swiss municipalities, p. 17

[3] *Bürgergemeinde* in German (http://www.hls-dhs-dss.ch/textes/d/D26443.php), French (http://www.hls-dhs-dss.ch/textes/f/F26443.php) and Italian (http://www.hls-dhs-dss.ch/textes/i/I26443.php) in the online *Historical Dictionary of Switzerland*.

[4] *Bäuert* in German (http://www.hls-dhs-dss.ch/textes/d/D10405.php), French (http://www.hls-dhs-dss.ch/textes/f/F10405.php) and Italian (http://www.hls-dhs-dss.ch/textes/i/I10405.php) in the online *Historical Dictionary of Switzerland*.

[5] *Degagna* in German (http://www.hls-dhs-dss.ch/textes/d/D35272.php), French (http://www.hls-dhs-dss.ch/textes/f/F35272.php) and Italian (http://www.hls-dhs-dss.ch/textes/i/I35272.php) in the online *Historical Dictionary of Switzerland*.

External links

- Amtliches Gemeindeverzeichnis der Schweiz/Liste officielle des communes de la Suisse/Elenco ufficiale dei Comuni della Svizzera, 2006 (http://www.bfs.admin.ch/bfs/portal/de/index/news/publikationen.html?publicationID=2321). Neuchâtel, 2006. ISBN 3-303-00334-3.
- *Nachbarschaft* in German (http://www.hls-dhs-dss.ch/textes/d/D16403.php), French (http://www.hls-dhs-dss.ch/textes/f/F16403.php) and Italian (http://www.hls-dhs-dss.ch/textes/i/I16403.php) in the online *Historical Dictionary of Switzerland*.

Sementina

Sementina	
Country Switzerland **Canton** Ticino **District** Bellinzona	
Population	3111 (Dec 2011)[1]
- Density	377 /km² (977 /sq mi)
Area	km² (sq mi)[2]
Elevation	224 m (ft)
Postal code	6514
SFOS number	5019
Surrounded by	Cugnasco, Giubiasco, Gudo, Monte Carasso
Website	www.sementina.ch [3] Profile [4] (Italian), SFSO statistics [5]

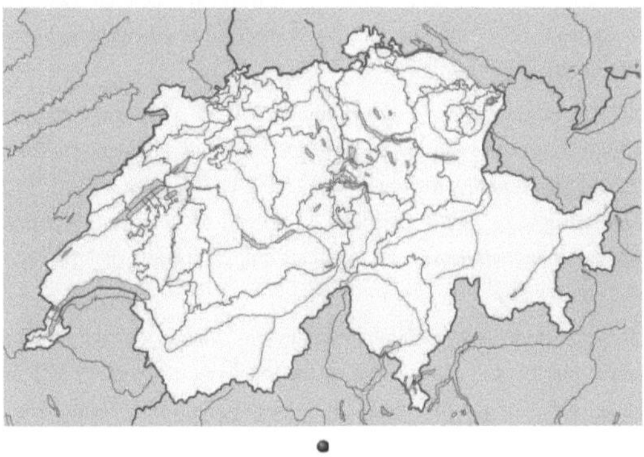

Sementina

Sementina is a municipality in the district of Bellinzona in the canton of Ticino in Switzerland.

Geography

Sementina has an area, as of 1997, of 8.25 square kilometers (sq mi). Of this area, 1.61 km² (sq mi) or 19.5% is used for agricultural purposes, while 5.56 km² (sq mi) or 67.4% is forested. Of the rest of the land, 0.84 km² (sq mi) or 10.2% is settled (buildings or roads), 0.21 km² (sq mi) or 2.5% is either rivers or lakes and 0.26 km² (sq mi) or 3.2% is unproductive land.

Of the built up area, housing and buildings made up 6.3% and transportation infrastructure made up 2.1%. Out of the forested land, 62.5% of the total land area is heavily forested and 3.4% is covered with orchards or small clusters of trees. Of the agricultural land, 8.0% is used for growing crops, while 5.2% is used for orchards or vine crops and 6.3% is used for alpine pastures. All the water in the municipality is flowing water. Of the unproductive areas, 2.2% is unproductive vegetation.[5]

Coat of arms

The blazon of the municipal coat of arms is *Per fess Gules a Wall embatteded throughout with a Towered Gate all Argent and Azure a Bar wavy Argent.*[6]

Demographics

Sementina has a population (as of December 2011) of 3111.[1] As of 2008, 24.7% of the population are foreign nationals.[7] Over the last 10 years (1997–2007) the population has changed at a rate of 15.4%.

Most of the population (as of 2000) speaks Italian(87.2%), with German being second most common (5.1%) and Serbo-Croatian being third (1.6%).[8] Of the Swiss national languages (as of 2000), 136 speak German, 39 people speak French, 2,306 people speak Italian, and 4 people speak Romansh. The remainder (161 people) speak another language.[9]

As of 2008, the gender distribution of the population was 49.0% male and 51.0% female. The population was made up of 1,090 Swiss men (35.8% of the population), and 401 (13.2%) non-Swiss men. There were 1,200 Swiss women (39.5%), and 350 (11.5%) non-Swiss women.[10]

In 2008 there were 19 live births to Swiss citizens and 4 births to non-Swiss citizens, and in same time span there were 11 deaths of Swiss citizens and 3 non-Swiss citizen deaths. Ignoring immigration and emigration, the population of Swiss citizens increased by 8 while the foreign population increased by 1. There were 5 Swiss men and 2 Swiss women who immigrated back to Switzerland. At the same time, there were 10 non-Swiss men and 17 non-Swiss women who immigrated from another country to Switzerland. The total Swiss population change in 2008 (from all sources) was an increase of 47 and the non-Swiss population change was a decrease of 3 people. This represents a population growth rate of 1.5%.[7]

The age distribution, as of 2009, in Sementina is; 329 children or 10.8% of the population are between 0 and 9 years old and 292 teenagers or 9.6% are between 10 and 19. Of the adult population, 326 people or 10.7% of the population are between 20 and 29 years old. 487 people or 16.0% are between 30 and 39, 512 people or 16.8% are between 40 and 49, and 348 people or 11.4% are between 50 and 59. The senior population distribution is 397 people or 13.1% of the population are between 60 and 69 years old, 215 people or 7.1% are between 70 and 79, there are 135 people or 4.4% who are over 80.[10]

As of 2000, there were 1,072 private households in the municipality, and an average of 2.4 persons per household.[8] In 2000 there were 504 single family homes (or 73.6% of the total) out of a total of 685 inhabited buildings. There were 75 two family buildings (10.9%) and 68 multi-family buildings (9.9%). There were also 38 buildings in the municipality that were multipurpose buildings (used for both housing and commercial or another purpose).[11]

The vacancy rate for the municipality, in 2008, was 1%. In 2000 there were 1,253 apartments in the municipality. The most common apartment size was the 4 room apartment of which there were 494. There were 51 single room apartments and 214 apartments with five or more rooms.[12] Of these apartments, a total of 1,062 apartments (84.8% of the total) were permanently occupied, while 155 apartments (12.4%) were seasonally occupied and 36 apartments (2.9%) were empty.[12] As of 2007, the construction rate of new housing units was 9.1 new units per 1000 residents.[8]

The historical population is given in the following table:[13]

year	population
1950	534
1960	618
1970	1,185
1980	2,122
1990	2,394
2000	2,646

Heritage sites of national significance

The Fortificazioni Ottocentesche (also known as the *Fortini Della Fame*, shared with Camorino and Monte Carasso) and the medieval settlement of S. Defendente are listed as Swiss heritage site of national significance.[14]

Politics

In the 2007 federal election the most popular party was the FDP which received 46.5% of the vote. The next three most popular parties were the CVP (18.2%), the SP (13.55%) and the Ticino League (10.3%). In the federal election, a total of 972 votes were cast, and the voter turnout was 55.7%.[15]

In the 2007 *Gran Consiglio* election, there were a total of 1,749 registered voters in Sementina, of which 1,329 or 76.0% voted. 22 blank ballots and 2 null ballots were cast, leaving 1,305 valid ballots in the election. The most popular party was the PLRT which received 508 or 38.9% of the vote. The next three most popular parties were; the SSI (with 230 or 17.6%), the PPD+GenGiova (with 192 or 14.7%) and the PS (with 160 or 12.3%).[16]

In the 2007 *Consiglio di Stato* election, there were 19 blank ballots and 4 null ballots, which left 1,306 valid ballots in the election. The most popular party was the PLRT which received 487 or 37.3% of the vote. The next three most popular parties were; the PPD (with 202 or 15.5%), the SSI (with 197 or 15.1%) and the LEGA (with 182 or 13.9%).[16]

Economy

As of 2007, Sementina had an unemployment rate of 3.98%. As of 2005, there were 61 people employed in the primary economic sector and about 16 businesses involved in this sector. 191 people are employed in the secondary sector and there are 26 businesses in this sector. 460 people are employed in the tertiary sector, with 73 businesses in this sector.[8] There were 1,286 residents of the municipality who were employed in some capacity, of which females made up 43.3% of the workforce.

In 2000, there were 528 workers who commuted into the municipality and 980 workers who commuted away. The municipality is a net exporter of workers, with about 1.9 workers leaving the municipality for every one entering. About 3.6% of the workforce coming into Sementina are coming from outside Switzerland.[17] Of the working population, 6.6% used public transportation to get to work, and 69.1% used a private car.[8]

As of 2009, there was one hotel in Sementina.[18]

Religion

From the 2000 census, 2,119 or 80.1% were Roman Catholic, while 156 or 5.9% belonged to the Swiss Reformed Church. There are 278 individuals (or about 10.51% of the population) who belong to another church (not listed on the census), and 93 individuals (or about 3.51% of the population) did not answer the question.[9]

Education

In Sementina about 65% of the population (between age 25-64) have completed either non-mandatory upper secondary education or additional higher education (either university or a *Fachhochschule*).[8]

In Sementina there are a total of 542 students (as of 2009). The Ticino education system provides up to three years of non-mandatory kindergarten and in Sementina there are 99 children in kindergarten. The primary school program lasts for five years and includes both a standard school and a special school. In the municipality, 176 students attend the standard primary schools and 13 students attend the special school. In the lower secondary school system, students either attend a two-year middle school followed by a two-year pre-apprenticeship or they attend a four-year program to prepare for higher education. There are 127 students in the two-year middle school and in their pre-apprenticeship, while 50 students are in the four-year advanced program.

The upper secondary school includes several options, but at the end of the upper secondary program, a student will be prepared to enter a trade or to continue on to a university or college. In Ticino, vocational students may either attend school while working on their internship or apprenticeship (which takes three or four years) or may attend school followed by an internship or apprenticeship (which takes one year as a full-time student or one and a half to two years as a part-time student).[19] There are 25 vocational students who are attending school full-time and 47 who attend part-time.

The professional program lasts three years and prepares a student for a job in engineering, nursing, computer science, business, tourism and similar fields. There are 5 students in the professional program.[20]

As of 2000, there were 94 students in Sementina who came from another municipality, while 244 residents attended schools outside the municipality.[17]

References

[1] Swiss Federal Statistics Office – STAT-TAB (http://www.pxweb.bfs.admin.ch/Dialog/varval.asp?ma=px-d-01-2A01&ti=Ständige+und+Nichtständige+Wohnbevölkerung+nach+Region,+Geschlecht,+Nationalität+und++Alter&path=../Database/German_01 - Bevölkerung/01.2 - Bevölkerungsstand und -bewegung/&lang=1&prod=01&openChild=true&secprod=2) *Ständige und Nichtständige Wohnbevölkerung nach Region, Geschlecht, Nationalität und Alter* (German) accessed 3 October 2012

[2] Arealstatistik Standard - Gemeindedaten nach 4 Hauptbereichen (http://www.bfs.admin.ch/bfs/portal/de/index/themen/02/03/blank/data/gemeindedaten.html)

[3] http://www.sementina.ch

[4] http://www.ti.ch/DFE/USTAT/DATI_COMUNI/0_pagine/default.asp?numero=5019

[5] Altitudine, superficie, secondo il genere di utilizzazione, rilevazione 1992/1997, e densità della popolazione, nel 2000 (http://www.ti.ch/DFE/USTAT/DATI_RIEPILOGHI/default_x.asp?menu=02&key1_testo=Territorio e ambiente&key1=02&key2=01&key3=01) (Italian) accessed 25 October 2010

[6] Flags of the World.com (http://www.fotw.us/flags/ch-ti019.html) accessed 2 November 2010

[7] Swiss Federal Statistical Office - Superweb database - Gemeinde Statistics 1981-2008 (http://www.bfs.admin.ch/bfs/portal/de/index/infothek/onlinedb/superweb/login.html) (German) accessed 19 June 2010

[8] Swiss Federal Statistical Office (http://www.bfs.admin.ch/bfs/portal/en/index/regionen/02/key.html) accessed 2 November 2010

[9] Popolazione residente, secondo la lingua principale e la religione, nel 2000 (http://www.ti.ch/DFE/USTAT/DATI_RIEPILOGHI/default_x.asp?menu=01&key1_testo=Popolazione&key1=01&key2=05&key3=00) (Italian) accessed 23 November 2010

[10] 01.02.03 Popolazione residente permanente (http://www.ti.ch/DFE/USTAT/DATI_RIEPILOGHI/default_x.asp?menu=01&key1_testo=Popolazione&key1=01&key2=02&key3=03) (Italian) accessed 23 November 2010

[11] 09.02.01 Edifici (http://www.ti.ch/DFE/USTAT/DATI_RIEPILOGHI/default_x.asp?menu=09&key1_testo=Costruzioni e abitazioni&key1=09&key2=02&key3=01) (Italian) accessed 23 November 2010

[12] 09.02.02 Abitazioni (http://www.ti.ch/DFE/USTAT/DATI_RIEPILOGHI/default_x.asp?menu=09&key1_testo=Costruzioni e abitazioni&key1=09&key2=02&key3=02) (Italian) accessed 23 November 2010

[13] Popolazione residente, dal 1850 (http://www.ti.ch/DFE/USTAT/DATI_RIEPILOGHI/default_x.asp?menu=01&key1_testo=Popolazione&key1=01&key2=02&key3=02) (Italian) accessed 23 November 2010

[15] Swiss Federal Statistical Office, *Nationalratswahlen 2007: Stärke der Parteien und Wahlbeteiligung, nach Gemeinden/Bezirk/Canton* (http://www.bfs.admin.ch/bfs/portal/de/index/themen/17/02/blank/data/04/03.html) (German) accessed 28 May 2010

[16] Elezioni cantonali: Gran Consiglio, Consiglio di Stato (http://www.ti.ch/DFE/USTAT/DATI_RIEPILOGHI/default_x.asp?menu=17&key1_testo=Politica&key1=17&key2=02&key3=02) (Italian) accessed 23 November 2010

[17] Swiss Federal Statistical Office - Statweb (http://www.media-stat.admin.ch/stat/pendler/pop.php) (German) accessed 24 June 2010

[18] Settori alberghiero e paralberghiero (http://www.ti.ch/DFE/USTAT/DATI_RIEPILOGHI/default_x.asp?menu=10&key1_testo=Turismo&key1=10&key2=03&key3=01) (Italian) accessed 23 November 2010

[19] EDK/CDIP/IDES (2010). KANTONALE SCHULSTRUKTUREN IN DER SCHWEIZ UND IM FÜRSTENTUM LIECHTENSTEIN / STRUCTURES SCOLAIRES CANTONALES EN SUISSE ET DANS LA PRINCIPAUTÉ DU LIECHTENSTEIN (http://edudoc.ch/record/35128/files/Schulsystem_alle.pdf) (Report). . Retrieved 24 June 2010.

[20] Allievi e studenti, secondo il genere di scuola, anno scolastico 2009/2010 (http://www.ti.ch/DFE/USTAT/DATI_RIEPILOGHI/default_x.asp?menu=15&key1_testo=Formazione e scienza&key1=15&key2=02&key3=01) (Italian) accessed 23 November 2010

Bellinzona

Bellinzona

Country	Switzerland
Canton	Ticino
District	Bellinzona

Population	17544 (Dec 2011)[1]
- Density	884 /km^2 (2290 /sq mi)
Area	km^2 (sq mi)[2]
Elevation	238 m (ft)
Postal code	6500
SFOS number	5002
Mayor	Mario Branda
Localities	Artore, Carasso , Daro, Ravecchia
Surrounded by	Arbedo-Castione, Giubiasco, Gorduno, Monte Carasso, Pianezzo, Sant'Antonio[3]
Website	www.bellinzona.ch [4] Profile [5] (Italian), SFSO statistics [5]

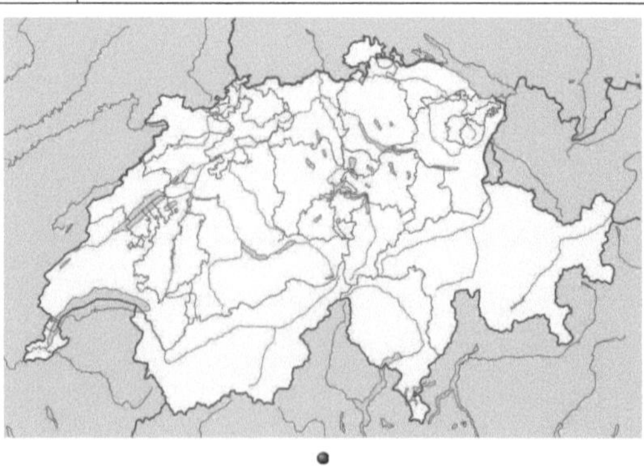

Bellinzona

Bellinzona (Italian pronunciation: [bɛllinˈdzoːna]; French: *Bellinzone*, pronounced: [bɛlɛ̃zon]; archaic German: *Bellenz* [ˈbɛlɛnts]; Latin *Bilitio* Latin pronunciation: [biˈlitsjo]) is the administrative capital of the canton Ticino in Switzerland. The city is famous for its three castles (Castelgrande, Montebello, Sasso Corbaro) that have been UNESCO World Heritage Sites since 2000.

The town lies east of the Ticino river, at the foot of the Alps. It stretches along the river valley, surrounded by the southern ranges of the Lepontine Alps to the east and west, and by the Lugano Prealps to the south.

Name and coat of arms

The toponym is first attested in 590 in Latin as *Belitio* or *Bilitio* (in the accusative, *Bilitionem*), by Gregory of Tours.[6] The name is Lepontic in origin, possibly from *belitio* ("juniper") or *belitione* ("juniper bushes").[7]

Coat of Arms

During the medieval period, the name is found as *Berinzona* (721, 762, 803, 1002), *Birrinzona* (1004), *Birizona* (1168), *Beliciona* (901, 977) and *Belinzona* (1055).[8] The German name of the town is *Bellenz*. A local folk etymology derives the name *Bellinzona* from *zona bellica* "war zone", making a connection to the Italian Wars.

The blazon of the municipal coat of arms is an erect serpent in silver on a red field. The fabulous animal is called in Italian "Biscione". This animal, which can also be found on the arms of the Alfa Romeo car company, is linked with the Visconti family, who were feudal lords of Bellinzona in the 14th and 15th centuries.[9] "

History

Pre-History and Roman era

Bellinzona has always occupied an important geographic location in the Alps. To the south, the Po valley is accessible by a lowland route down the valley of the Ticino river and by Lake Maggiore. To the north, the valley of the Ticino leads to the high alpine passes of Nufenen, St. Gotthard, Lukmanier and San Bernardino. Although now little used, the San Jorio Pass to the east was also important in Bellinzona's past.[10]

While the region has been occupied since the early Neolithic age[10] [11] it wasn't until the late 1st century BC that a fort was built in the area during the reign of the Roman Emperor Augustus. While the fort fell into disrepair in the following centuries, it was rebuilt and greatly expanded in the 4th century AD. During the reign of Diocletian and Constantin a chain of castles and watchtowers were built to protect northern Italy from invasion. Bellinzona's location was recognized as a key point in the defenses and a large castle was built to protect the walls. The town that grew up around the fortifications was known as *Bilitio*.

Following the collapse of the Western Roman Empire the successor states, which included the Ostrogoths around 500 AD, the eastern Byzantine Empire towards the middle of the 6th century, and the Longobards from 568/70, all took control of Bellinzona and used the castle to assert control of the surrounding passes. Under the Longobards, Bellinzona became the site of a permanent garrison to protect the region from raids by the neighboring Frankish and Alemannic tribes.[12] From Bellinzona the Longobards controlled the traffic on the important trade route from Varese over Ponte Tresa, the Monte Ceneri Pass, Biasca and finally over the Lukmanier Pass into Chur. Some researchers believe that Bellinzona may have been the capital of a county that included most of the valleys in Ticino.[12]

Early Middle Ages

At around 774 the Frankish Kingdom (that would become the Carolingian Empire) gained control of the Ticino valley including Bellinzona.

About two centuries later the Holy Roman Emperor Otto III, seeking to restore the power of glory of ancient Rome and expand into Italy, opened the Lukmanier and St. Bernard passes. Control of Bellinzona was a key part of this expansion. The city was taken from Milan and given as a gift to the Bishop of Como, who supported the Ottonian dynasty. In 1002, following the death of Otto III, Marquis Arduino of Ivrea declared himself King of Italy and ratified the bishop's ownership of the *Castelgrande* and the city. Two years later, after Arduino had been defeated by Henry II the King of Germany, Henry II's man Enrico II reratified the gift of the *Castlegrande* on the Bishop of Como.[] The city is mentioned in medieval sources in 1218 as *Bilizione*.

Conflicts between the Pope and the Emperor

Coat of arms of the House of Visconti. Under the Visconti, Bellinzona flourished and the city was expanded.

During the Investiture Controversy of the late 11th century the city of Bellinzona with its castle came under the control of the Hohenstaufens of Swabia. However, in 1180, Frederick I (Barbarossa) placed the city under the jurisdiction of the city of Como.[] In the following years Como tended to support the Pope in his conflicts with the Holy Roman Emperor. However in 1239, Como sided with the Emperor Frederick II who quickly moved forces into Bellinzona and strengthened the *Castelgrande*. In 1242 Milan sent Guelph (or pro-papacy) forces under the command of Simone di Orello to take Bellinzona.[] The city and castle were taken which weakened the Emperor south of the Alps. However the town was back under the jurisdiction of Como in 1249.[12] Conflicts in northern Italy continued, the *Castelgrande* was besieged several times in 1284, 1292 and 1303. During this time the Rusca family in Como, a Ghibelline or pro-Imperial family, fought the growing power of Milan under the pro-papacy House of Visconti with limited success. Around the end of the 13th century the Rusca family built another castle, *Montebello*, in Bellinzona, which they controlled. This was fortunate because by 1335 the Rusca family had been driven out of Como and had to retreat to Bellinzona. Five years later, in 1340, Milan besieged Bellinzona. Following a lengthly siege, the city fell to Milan but the Ruscas were allowed to keep *Montebello*.[] Pro-papacy Milan would dominate Bellinzona for the next one and a half centuries, though the pro-Imperial Rusca would also occupy part of the city.

Expansion of Bellinzona under Milan

The *Murata* or city wall of Bellinzona

Under the control of the Visconti trade flourished and the city of Bellinzona grew. Even when an alternative route over the Alps, the Schöllenen bridge opened, traffic in the St. Gotthard increased to the highest levels ever.[13] During the second half of the 14th century a long wall, known as the *Murata*, was built across the entire Tessin valley. This wall allowed Milan to protect and tax the trade route over the St. Gotthard Pass.[12] While the city was controlled by Milan through the Visconti after 1340, the Visconti did not have a formal title and feudal rights until 1396 when they were granted by King Wenceslaus. However, the orderly growth of Bellinzona was threatened in 1402 when Duke Gian Galeazzo Visconti died. In 1403 Bellinzona was taken under the control of Alberto di Sacco of Val Mesolcina. He held Bellinzona until 1419 when it was taken over by Uri and Obwalden which had expanded into the Leventina Valley. Milan attacked the city three years later in 1422 after an offer to buy the city was rejected by the Swiss Confederation. The troops from Uri and Obwalden were quickly driven from the city and later defeated at the Battle of Arbedo on 30 June 1422. This defeat discouraged the expansionist intentions of Uri and its allies towards Lake Maggiore for a time.

During the period of unrest following Gian Galeazzo Visconti's death, a tower which would become the nucleus of the third castle, *Sasso Corbaro*, was built outside the city.

While the border between Uri and Milan was fixed in the peace treaty of 1426, in 1439 Uri invaded again. While they were unable to take Bellinzona, the victories of the Swiss troops led to Milan granting all of the Leventina Valley to Pollegio to Uri in 1441. Following the death of Duke Filippo Maria Visconti in 1447, Bellinzona was in the middle of the succession crisis between Franchino Rusca of Locarno and Heinrich of Val Mesolcina, who were allied with Uri and the Ambrosian Republic in Milan. The war following the succession crisis lasted nearly three years until Francesco I Sforza seized power in Milan. Bellinzona quickly accepted the new Sforza dynasty and the peace and stability that followed.[12]

The peace was broken again in 1478 when the Swiss once again attacked Bellinzona unsuccessfully. However Swiss pride was restored by the Battle of Giornico which followed, where a force of 600 Swiss soldiers defeated 10,000 Milanese troops. Following the attack, Milan built the *Sasso Corbaro* either on the site of a tower which had been built nearly a century before.[12] The other two castles were strengthened and the *Murata* wall across the valley was rebuilt. Much of the modern castles and fortifications date from this period of construction in the late 15th century.

An associate of the Swiss Confederation

Castles of Bellinzona

In 1499 nearly one and a half centuries of Milanese rule ended with the invasion of Milan by Louis XII of France. He captured Bellinzona and fearing an attack by the Swiss, fortified the *Castelgrande* with 1000 troops.[14] Throughout the winter of 1499/1500 unrest in Bellinzona grew, until January when an armed revolt of the citizens of Bellinzona drove the French troops from the city. Following the capture and execution of Ludovico Sforza in April 1500 and seeking protection from France, Bellinzona joined the Swiss Confederation on 14 April 1500, as a condominium under the joint administration of Uri, Schwyz and Nidwalden. Following the Napoleonic invasion of Switzerland in 1798, Bellinzona was the capital of the canton of Bellinzona within the Helvetic Republic (1798–1803).

Bellinzona since 1803

Following the Act of Mediation in 1803 Bellinzona became part of the independent canton of Ticino, and the capital of the new canton from 1803 to 1814. From that date until 1878, Bellinzona, Lugano, and Locarno, took turns being capital every six years.

Train disaster in Bellinzona, May 1924

The city includes the village of Artore and, since the incorporation in 1907, the former municipalities of Carasso, Daro, and Ravecchia.

In 1874, the first sections of the Gotthard railway opened, linking Bellinzona to Biasca and Locarno. By 1882, the whole line was open, extending northwards to northern Switzerland via the Gotthard Tunnel, southwards to Lugano and Milan via the Monte Ceneri Pass, and down the east shore of Lake Maggiore to Luino. Between 1907 and 1972, Bellinzona was also linked to Mesocco and other Val Mesolcina communities by the Bellinzona–Mesocco railway.

Geography

Bellinzona is situated in the valley of the Ticino river, at an altitude of 238 metres (ft). The city centre lies about 1 kilometre (mi) east of the river, with the urban area confined to the bottom and lower slopes of the valley. However the city's boundaries extend up both sides of the valley, to altitudes of 1850 metres (ft) to the west, and 1700 metres (ft) to the east.[3]

View of Bellinzona from Castello Montebello

The city is located at a point at which the Ticino river, which has been flowing in a generally southerly direction, makes a turn to the east, to flow through Lake Maggiore to the Po Valley and Lombardy. Upstream the valley reaches into the high alps, with access to the passes of Nufenen, St. Gotthard and Lukmanier. The Mesolcina valley, which leads to the San Bernardino Pass, joins the Ticino on the northern boundary of the city. To the south the Monte Ceneri Pass crosses the Lugano Prealps to give access to Lake Lugano and an alternate route to Lombardy, whilst the San Jorio Pass provides a possible route east to Lake Como.[3] [10]

Piazza del Sole, Post Office and Station

Bellinzona has an area, as of 1997, of 19.15 square kilometers (sq mi). Of this area, 4.21 km^2 (sq mi) or 22.0% is used for agricultural purposes, while 11.96 km^2 (sq mi) or 62.5% is forested. Of the rest of the land, 4.55 km^2 (sq mi) or 23.8% is settled (buildings or roads), 0.47 km^2 (sq mi) or 2.5% is either rivers or lakes and 0.09 km^2 (acres) or 0.5% is unproductive land.[15]

Of the built up area, industrial buildings made up 1.6% of the total area while housing and buildings made up 13.0%. Transportation infrastructure made up 5.8% while parks, green belts and sports fields made up 2.7%. Out of the forested land, 60.1% of the total land area is heavily forested and 1.9% is covered with orchards or small clusters of trees. Of the agricultural land, 6.6% is used for growing crops, while 2.5% is used for orchards or vine crops and 13.0% is used for alpine pastures. All the water in the municipality is flowing water.[15]

Demographics

Bellinzona has a permanent population (as of December 2011) of 17544[1] . In 2008, 29.4% of the population were foreign nationals.[16] Between 1997 and 2007 the population has changed at a rate of 0.7%.

Most of the population (as of 2000) speaks Italian (87.4%), with German being second most common (3.6%) and Serbo-Croatian being third (2.5%).[17] Of the Swiss national languages (as of 2000), 14,392 people speak Italian, 590 speak German, 189 people speak French, and 13 people speak Romansh. The remainder (1,279 people) speak another language.[18] The metropolitan area of Bellinzona had a population of 47,128,[] divided into 16 municipalities.

As of 2008, the gender distribution of the population was 46.6% male and 53.4% female. The population was made up of 5,503 Swiss men (31.8% of the population), and 2,567 (14.8%) non-Swiss men. There were 6,781 Swiss women (39.1%), and 2,472 (14.3%) non-Swiss women.[19]

In 2008 there were 132 live births to Swiss citizens and 45 births to non-Swiss citizens, and in same time span there were 132 deaths of Swiss citizens and 15 non-Swiss citizen deaths. Ignoring immigration and emigration, the population of Swiss citizens remained the same while the foreign population increased by 30. There were 7 Swiss men who emigrated from Switzerland to another country, 3 Swiss women who emigrated from Switzerland to another country, 67 non-Swiss men who emigrated from Switzerland to another country and 70 non-Swiss women who emigrated from Switzerland to another country. The total Swiss population change in 2008 (from all sources) was an increase of 377 and the non-Swiss population change was a decrease of 202 people. This represents a population growth rate of 1.0%.[16]

The age distribution, as of 2009, in Bellinzona is; 1,530 children or 8.8% of the population are between 0 and 9 years old and 1,623 teenagers or 9.4% are between 10 and 19. Of the adult population, 2,091 people or 12.1% of the population are between 20 and 29 years old. 2,526 people or 14.6% are between 30 and 39, 2,721 people or 15.7% are between 40 and 49, and 2,260 people or 13.0% are between 50 and 59. The senior population distribution is 1,969 people or 11.4% of the population are between 60 and 69 years old, 1,470 people or 8.5% are between 70 and 79, there are 1,133 people or 6.5% who are between 80 and 89.[19]

As of 2000, there were 7,294 private households in the municipality, and an average of 2.2 persons per household.[17] In 2000 there were 1,490 single family homes (or 51.5% of the total) out of a total of 2,892 inhabited buildings. There were 419 two family buildings (14.5%) and 642 multi-family buildings (22.2%). There were also 341 buildings in the municipality that were multipurpose buildings (used for both housing and commercial or another purpose).[20]

The vacancy rate for the municipality, in 2008, was 1.45%. Of the apartments, a total of 7,255 apartments (85.8% of the total) were permanently occupied, while 932 apartments (11.0%) were seasonally occupied and 268 apartments (3.2%) were empty.[21] In 2000 there were 8,455 apartments in the municipality. The most common apartment size was the 3 room apartment of which there were 2,746. There were 474 single room apartments and 1,253 apartments with five or more rooms.[21] As of 2007, the construction rate of new housing units was 6.2 new units per 1000 residents.[17]

As of 2003 the average price to rent an average apartment in Bellinzona was 956.03 Swiss francs (CHF) per month (US$760, £430, €610 approx. exchange rate from 2003). The average rate for a one room apartment was 673.24 CHF (US$540, £300, €430), a two room apartment was about 740.60 CHF (US$590, £330, €470), a three room apartment was about 910.37 CHF (US$730, £410, €580) and a six or more room apartment cost an average of 1406.75 CHF (US$1130, £630, €900). The average apartment price in Bellinzona was 85.7% of the national average of 1116 CHF.[22]

Historic Demographics

		1591	1781	1808	1850	1880	1910	1930	1950	1970	1990
Population		ca. 200[A]	ca. 1,100	1,260	3,209	4,036	10,406	10,706	12,060	16,979	16,849
Language	German					140	1,028	831	807	1,040	681
	French					6	74	127	162	179	209
	Italian					3,887	9,266	9,712	11,053	15,574	14,948
	Other					3	38	36	38	186	1,011
Religion	Protestant					43	632	550	577	844	626
	Roman Catholic					3,985	8,947	9,577	11,196	15,817	14,592
	Other/None					8	827	579	287	318	1,631[B]
Nationality	Swiss				2,742	3,260	6,936	8,755	10,427	12,848	11,924
	Foreign				467	776	3,470	1,951	1,633	4,131	4,925

A Number of households

B in 1990, 879 were either atheist or did not identify with any religion

Elections

In the 2007 federal election the most popular party was the FDP which received 30.61% of the vote. The next three most popular parties were the SP (23.11%), the CVP (19.85%) and the Ticino League (11.42%). In the federal election, a total of 4,634 votes were cast, and the voter turnout was 46.1%.[23]

In the 2007 *Ticino Gran Consiglio* election, there were a total of 10,187 registered voters in Bellinzona, of which 6,486 or 63.7% voted. 109 blank ballots and 16 null ballots were cast, leaving 6,361 valid ballots in the election. The most popular party was the PLRT which received 1,569 or 24.7% of the vote. The next three most popular parties were; the SSI (with 1,233 or 19.4%), the PS (with 1,210 or 19.0%) and the PPD+GenGiova (with 957 or 15.0%).[24]

In the 2007 *Ticino Consiglio di Stato* election, there were 60 blank ballots and 22 null ballots, which left 6,405 valid ballots in the election. The most popular party was the PS which received 1,472 or 23.% of the vote. The next three most popular parties were; the PLRT (with 1,453 or 22.7%), the SSI (with 1,103 or 17.2%) and the LEGA (with 1,074 or 16.8%).[24]

Economy

As of 2007, Bellinzona had an unemployment rate of 5.16%. As of 2005, there were 33 people employed in the primary economic sector and about 10 businesses involved in this sector. 1,691 people are employed in the secondary sector and there are 149 businesses in this sector. 11,647 people are employed in the tertiary sector, with 1,093 businesses in this sector.[17]

Office building originally built for the Directorate of Telecommunications of the former Swiss PTT, now a business center.

In 2000, there were 16,293 workers who commuted into the municipality and 2,631 workers who commuted away. The municipality is a net importer of workers, with about 6.2 workers entering the municipality for every one leaving. About 45.3% of the workforce coming into Bellinzona are coming from outside Switzerland, while 0.1% of the locals commute out of Switzerland for work.[25] Of the working population, 8.8% used public transportation to get to work, and 50.6% used a private car.[17]

As of 2009, there were 9 hotels in Bellinzona with a total of 145 rooms and 283 beds.[26]

Religion

From the 2000 census, 12,185 or 74.0% were Roman Catholic, while 651 or 4.0% belonged to the Swiss Reformed Church. There are 2,164 individuals (or about 13.14% of the population) who belong to another church (not listed on the census), and 1,463 individuals (or about 8.89% of the population) did not answer the question.[18]

Swiss Reformed church in Bellinzona

Weather

Bellinzona has an average of 102.8 days of rain or snow per year and on average receives 1563 mm (in) of precipitation. The wettest month is May during which time Bellinzona receives an average of 181 mm (in) of rain or snow. During this month there is precipitation for an average of 13 days. The driest month of the year is December with an average of 60 mm (in) of precipitation over 13 days.[27]

Education

In Bellinzona about 60.5% of the population (between age 25–64) have completed either non-mandatory upper secondary education or additional higher education (either university or a *Fachhochschule*).[17]

In Bellinzona there are a total of 2,662 students (as of 2009). The Ticino education system provides up to three years of non-mandatory kindergarten and in Bellinzona there are 413 children in kindergarten.

The primary school program lasts for five years and includes both a standard school and a special school. In the municipality, 781 students attend the standard primary schools and 51 students attend the special school. In the lower secondary school system, students either attend a two-year middle school followed by a two-year pre-apprenticeship or they attend a four-year program to prepare for higher education. There are 632 students in the two-year middle school and 3 in their pre-apprenticeship, while 271 students are in the four-year advanced program.

The upper secondary school includes several options, but at the end of the upper secondary program, a student will be prepared to enter a trade or to continue on to a university or college. In Ticino, vocational students may either attend school while working on their internship or apprenticeship (which takes three or four years) or may attend school followed by an internship or apprenticeship (which takes one year as a full-time student or one and a half to two years as a part-time student).[28] There are 162 vocational students who are attending school full-time and 299 who attend part-time. The professional program lasts three years and prepares a student for a job in engineering, nursing, computer science, business, tourism and similar fields. There are 50 students in the professional program.[29]

As of 2000, there were 2,957 students in Bellinzona who came from another municipality, while 313 residents attended schools outside the municipality.[25]

Bellinzona is home to 2 libraries. These libraries include; the Biblioteca Cantonale Bellinzona and the *Biblioteca comunale*. There was a combined total (as of 2008) of 138,818 books or other media in the libraries, and in the same year a total of 43,919 items were loaned out.[30]

Transportation

Autopostale city bus (left) and longer distance bus in Bellinzona

Bellinzona railway station is a major intermediate stop on the Gotthard railway. It is a stopping point for major trains heading north toward Arth-Goldau and Zürich, south toward Lugano, Chiasso and Italy, or southwest to Locarno. The station is also served by the regional trains operated by TiLo to Biasca, Chiasso, Locarno, Lugano and Malpensa Airport.

PostBus Switzerland, known locally as the *AutoPostale*, operate a small network of city bus routes within Lugano, as well as longer distance routes to other towns and cities. All routes serve the railway station.

The A2 and A13 motorways, as well as some main roads, link here, thus making it an important transportational node. The A2 runs north via the Gotthard Pass to Lucerne, Basle and Germany, and south to Lugano and Italy. The A13 runs north-east via the San Bernardino Pass to Chur and Austria.

Culture

The city is known for its carnival Rabadan, which has taken place for over 150 years.

Sport

The football club AC Bellinzona play in the Swiss Challenge League. Their stadium is the Stadio Comunale.

Heritage sites of national significance

Bellinzona is home to twelve buildings or areas that are listed as Swiss heritage site of national significance. Additionally, it is home to the UNESCO World Heritage Site of the Three Castles of Bellinzona. Finally, entire old city of Bellinzona is listed on the Inventory of Swiss Heritage Sites.[31]

In addition to the Three Castles and the city walls, the medieval and early modern city is included on the list. Three religious buildings, the Church of S. Maria delle Grazie, the Collegiata dei Ss. Pietro e Stefano and the Church of S. Biagio a Ravecchiai are on the list. The Cantonal Archives, *Bagno Pubblico*, the secondary school (Italian: *Scuola media*) on via Lavizzari 28 and the *Teatro sociale* are the rest of the buildings on the list.

Castelgrande

Montebello

Castello di Sasso Corbaro

City Wall

Church di S. Maria delle Grazie

Collegiata dei Ss. Pietro e Stefano

Church of S. Biagio a Ravecchia

Old city of Bellinzona

Bellinzona by J. M. W. Turner, 1841.

The Three Castles

Ramparts of Bellizona connecting Castelgrande to Castello di Montebello

The Three Castles, officially listed as the **Three Castles, Defensive Wall and Ramparts of the Market-Town of Bellinzone**, have been an UNESCO World Heritage Site since 2000. The group is composed of *Castelgrande*, castle *Montebello*, castle *Sasso Corbaro* and fortified walls. The *Castelgrande* is located on a rocky peak overlooking the valley, with a series of fortified walls that protect the old city and connect to the *Montebello*. The third castle (*Sasso Corbaro*) is located on an isolated rocky promontory south-east of the other two.[32]

Castelgrande

Castelgrande showing the walls and towers of the extensive castle

The site of the *Castelgrande* has been fortified since at least the late 1st century BC and until the 13th century it was the only fortification in Bellinzona. During its history the castle has been known as the stronghold (before the 13th century), the Old Castle in the 14–15th centuries, *Un Castle* after 1506 and Saint Michael's Castle from 1818.[]

The *Castelgrande* hill includes a nearly vertical side on the north and a steep southern side, but is nearly flat and 150-200m in diameter. The natural shape of the hill has encouraged every man-made fortification to follow the same contours. While the Roman fort is not visible the Roman foundations were used by the High Middle Ages castle which followed. Of the High Middle Ages castle the only visible parts are a few pieces of wall that are still standing. Much of the visible castle dates from 1250–1500 with extensive renovations and some expansion in the last two centuries. Most of the area inside the castle walls is now flat, open space.

Records from the 11th to 15th centuries as well as archeological evidence indicate that the castle grounds were once full of buildings. However most of these were pulled down by the Dukes of Milan to free up interior space. The open space was divided into 3 large baileys which served to provide temporary housing for troops that could be stationed in Bellinzona. Under the Dukes of Milan the outer fortifications were extended and strengthened. The walls were raised, extended and towers were added. The western walls were totally rebuilt and connected to the city walls.

Montebello castle located on a rocky hilltop east of town is connected to *Castelgrande* by the city walls

The castle can be reached by taking an elevator from the foot of the rock to the castle grounds or by climbing steep, narrow streets from the city center through the city wall onto the castle grounds.

Montebello

Sasso Corbaro castle

Montebello Castle (known as the Small, New or Middle Castle in the 15th century, as Schwyz Castle from 1506 and St. Martin's Castle after 1818) is located to the east of the town center. It was built before 1313 for the pro-Imperial Rusca family, who occupied the castle following the Visconti victory and occupation of the *Castelgrande*. By the end of the 14th century it was in the hands of the Visconti.[33] The castle was renovated and expanded between 1462 and 1490 to its current state. In the 19th century the castle fell into disrepair and was renovated starting in 1903.

A little chapel, dedicated to Saint Michael, leans against the wall of the more recent south-facing section; built around 1600, it is one of the few buildings erected in the castles of Bellinzona under the rule of the three Swiss cantons.

Montebello Castle houses the Archaeological and Civic Museum. The museum was opened in 1974 and is located in the tower and the former residential quarters of Montebello Castle. It is divided into two sections-history and archaeology. In the history section there are several capitals from the 15th century and a rare 13th-century Baptismal font as well as drawings and sketchs from several artists. This section also houses a collection of ceremonial and military arms. The archaeology section includes many items from 1400–1500 BC as well as ceramics, glassware, funeral urns, ornamental objects and jewellery in iron and bronze from around the canton. The museum is open from March to November.[34]

Sasso Corbaro

Sasso Corbaro, known as Unterwalden Castle after 1506 and Saint Barbara's Castle after 1818, is about 600m south-west of town on a rocky hill. Unlike the other two castles Sasso Corbaro is not integrated into the city walls. The first part of the castle was the north-eastern tower which was built in 1478 to close a gap in the defenses of the city.[35] The walls and south-west tower were added later. The castle was struck by lightning multiple times during the 16th and 17th centuries, and by 1900 was falling into ruins.

Today, Sasso Corbaro Castle houses the *Sala Emma Poglia* which is the "wooden room" built for the Emma family during the 17th century. Originally located in the entrance hall of their home in Olivone in the Blenio Valley, the room was purchased by the Canton of Ticino in 1944 and housed first in the *Castelgrande* before being moved to the Sasso Corbaro in 1989. The room is panelled entirely in walnut and also includes the *stüva*, stove which provided heating. The stove bears the crest of the Emma family (an eagle and a lion rampant). The museum also houses temporary exhibits. It is open from March until November.[36]

References

[1] Swiss Federal Statistics Office – STAT-TAB (http://www.pxweb.bfs.admin.ch/Dialog/varval.asp?ma=px-d-01-2A01&ti=Ständige+und+Nichtständige+Wohnbevölkerung+nach+Region,+Geschlecht,+Nationalität+und++Alter&path=../Database/German_01 - Bevölkerung/01.2 - Bevölkerungsstand und -bewegung/&lang=1&prod=01&openChild=true&secprod=2) *Ständige und Nichtständige Wohnbevölkerung nach Region, Geschlecht, Nationalität und Alter* (German) accessed 3 October 2012

[2] Arealstatistik Standard - Gemeindedaten nach 4 Hauptbereichen (http://www.bfs.admin.ch/bfs/portal/de/index/themen/02/03/blank/data/gemeindedaten.html)

[3] Swiss Confederation. *map.geo.admin.ch* (http://map.geo.admin.ch/?selectedNode=node_ch.swisstopo.swissboundaries3d-gemeinde-flaeche.fill1&Y=722582.5&X=117523.5&zoom=6&bgLayer=ch.swisstopo.pixelkarte-farbe&layers=ch.swisstopo.swissboundaries3d-gemeinde-flaeche.fill&layers_opacity=1&layers_visibility=true&lang=en) (Map). . Retrieved 2012-11-17.

[4] http://www.bellinzona.ch

[5] http://www.ti.ch/DFE/USTAT/DATI_COMUNI/0_pagine/default.asp?numero=5002

[6] Gregorius Turonensis, Historiae, 10.3 (http://monumenta.ch/latein/text.php?tabelle=Gregorius_Turonensis&rumpfid=Gregorius Turonensis, Historiae, 10, 3&domain=&lang=0&links=&inframe=1&hide_apparatus=1&PHPSESSID=05b6e09fa203c1964e1b888c9ee7d12e), *Quod exercitus Childeberthi regis in Italiam abiit: Olo autem dux ad Bilitionem huius urbis castrum, in campis situm Caninis, inportunae accedens, iaculo sub papilla sauciatus, cecidit et mortuus est* "Duke Olo went rashly to *Bilitio*, a stronghold of this city [Milan], situated on the plains called Canini, and was wounded with a dart under the nipple and fell and died."

[7] *Archiv für das Studium der neueren Sprachen und Literaturen* 184, J. Bädeker, 1943, p. 105 (http://books.google.ch/books?id=FVo4AQAAIAAJ&q=Belitio++Bellinzona&dq=Belitio++Bellinzona&hl=de&ei=p9b1TdOGEoaDOoqk5KoH&sa=X&oi=book_result&ct=result&resnum=3&ved=0CDYQ6AEwAg). Jacob Früh, *Geographie der Schweiz*, Fehr, 1932, p. 12.

[8] *Anzeiger für schweizerische Altertumskunde, Bände 21–24*, J. Herzog, 1912, p. 404.

[9] Pascal Gross, 11 January 2003 at Flags of the World.com (http://www.fotw.us/flags/ch-ti002.html) accessed 25-October-2010

[10] *Bellinzona-Ancient and early history* in German (http://www.hls-dhs-dss.ch/textes/d/D2031-1-1.php), French (http://www.hls-dhs-dss.ch/textes/f/F2031-1-1.php) and Italian (http://www.hls-dhs-dss.ch/textes/i/I2031-1-1.php) in the online *Historical Dictionary of Switzerland.*

[11] Official Website-Prehistoric Settlement (http://www.bellinzonaunesco.ch/castelliunesco/en/la_storia/colonizzazione_preistorica.asp) accessed 7 July 2008

[12] *Bellinzona-The Middle Ages* in German (http://www.hls-dhs-dss.ch/textes/d/D2031-1-3.php), French (http://www.hls-dhs-dss.ch/textes/f/F2031-1-3.php) and Italian (http://www.hls-dhs-dss.ch/textes/i/I2031-1-3.php) in the online *Historical Dictionary of Switzerland.*

[13] Official Site-Late Middle Ages (http://www.bellinzonaunesco.ch/castelliunesco/en/la_storia/tardo_medioevo.asp) accessed 17 July 2008

[14] Official Site-Bellinzona joins the Confederation (http://www.bellinzonaunesco.ch/castelliunesco/en/la_storia/bellinzona_confederata.asp) accessed 17 July 2008

[15] Altitudine, superficie, secondo il genere di utilizzazione, rilevazione 1992/1997, e densità della popolazione, nel 2000 (http://www.ti.ch/DFE/USTAT/DATI_RIEPILOGHI/default_x.asp?menu=02&key1_testo=Territorio e ambiente&key1=02&key2=01&key3=01) (Italian) accessed 25 October 2010

[16] Swiss Federal Statistical Office – Superweb database – Gemeinde Statistics 1981–2008 (http://www.bfs.admin.ch/bfs/portal/de/index/infothek/onlinedb/superweb/login.html) (German) accessed 19 June 2010

[17] Swiss Federal Statistical Office (http://www.bfs.admin.ch/bfs/portal/en/index/regionen/02/key.html) accessed 25-October-2010

[18] Popolazione residente, secondo la lingua principale e la religione, nel 2000 (http://www.ti.ch/DFE/USTAT/DATI_RIEPILOGHI/default_x.asp?menu=01&key1_testo=Popolazione&key1=01&key2=05&key3=00) (Italian) accessed 23 November 2010

[19] 01.02.03 Popolazione residente permanente (http://www.ti.ch/DFE/USTAT/DATI_RIEPILOGHI/default_x.asp?menu=01&key1_testo=Popolazione&key1=01&key2=02&key3=03) (Italian) accessed 23 November 2010

[20] 09.02.01 Edifici (http://www.ti.ch/DFE/USTAT/DATI_RIEPILOGHI/default_x.asp?menu=09&key1_testo=Costruzioni e abitazioni&key1=09&key2=02&key3=01) (Italian) accessed 23 November 2010

[21] 09.02.02 Abitazioni (http://www.ti.ch/DFE/USTAT/DATI_RIEPILOGHI/default_x.asp?menu=09&key1_testo=Costruzioni e abitazioni&key1=09&key2=02&key3=02) (Italian) accessed 23 November 2010

[22] Swiss Federal Statistical Office-Rental prices (http://www.bfs.admin.ch/bfs/portal/de/index/themen/05/06/blank/key/einfuehrung.html) 2003 data (German) accessed 26 May 2010

[23] Swiss Federal Statistical Office, *Nationalratswahlen 2007: Stärke der Parteien und Wahlbeteiligung, nach Gemeinden/Bezirk/Canton* (http://www.bfs.admin.ch/bfs/portal/de/index/themen/17/02/blank/data/04/03.html) (German) accessed 28 May 2010

[24] Elezioni cantonali: Gran Consiglio, Consiglio di Stato (http://www.ti.ch/DFE/USTAT/DATI_RIEPILOGHI/default_x.asp?menu=17&key1_testo=Politica&key1=17&key2=02&key3=02) (Italian) accessed 23 November 2010

[25] Swiss Federal Statistical Office – Statweb (http://www.media-stat.admin.ch/stat/pendler/pop.php) (German) accessed 24 June 2010

[26] Settori alberghiero e paralberghiero (http://www.ti.ch/DFE/USTAT/DATI_RIEPILOGHI/default_x.asp?menu=10&key1_testo=Turismo&key1=10&key2=03&key3=01) (Italian) accessed 23 November 2010
[27] , the weather station elevation is 225 meters above sea level.
[28] EDK/CDIP/IDES (2010). KANTONALE SCHULSTRUKTUREN IN DER SCHWEIZ UND IM FÜRSTENTUM LIECHTENSTEIN / STRUCTURES SCOLAIRES CANTONALES EN SUISSE ET DANS LA PRINCIPAUTÉ DU LIECHTENSTEIN (http://edudoc.ch/record/35128/files/Schulsystem_alle.pdf) (Report). . Retrieved 24 June 2010.
[29] Allievi e studenti, secondo il genere di scuola, anno scolastico 2009/2010 (http://www.ti.ch/DFE/USTAT/DATI_RIEPILOGHI/default_x.asp?menu=15&key1_testo=Formazione e scienza&key1=15&key2=02&key3=01) (Italian) accessed 23 November 2010
[30] Swiss Federal Statistical Office, list of libraries (http://www.bfs.admin.ch/bfs/portal/de/index/themen/16/02/02/data.html) (German) accessed 14 May 2010
[33] Official Website-Montebello Castle (http://www.bellinzonaunesco.ch/castelliunesco/en/castelli_e_mura/montebello.asp) accessed 30 July 2008
[34] Bellinzona Tourism Website-Montebello Castle Museum (http://www.bellinzonaturismo.ch/framework/Desktopdefault.aspx?menu_id=373) accessed 30 July 2008
[35] Official Website-Sasso Corbaro (http://www.bellinzonaunesco.ch/castelliunesco/en/castelli_e_mura/sasso_corbaro.asp) accessed 31 July 2008
[36] Bellinzona Tourism-Sasso Corbaro (http://www.bellinzonaturismo.ch/framework/Desktopdefault.aspx?menu_id=374) accessed 31 July 2008

- *Bellinzona (municipality)* in German (http://www.hls-dhs-dss.ch/textes/d/D2031.php), French (http://www.hls-dhs-dss.ch/textes/f/F2031.php) and Italian (http://www.hls-dhs-dss.ch/textes/i/I2031.php) in the online *Historical Dictionary of Switzerland*.

External links

- Bellinzona official website (http://www.bellinzona.ch/)
- Rabadan carnival (http://www.rabadan.ch)
- Unesco (http://www.bellinzonaunesco.ch/castelliunesco/en/)

Kingdom_of_Lombardy–Venetia

Kingdom of Lombardy–Venetia
***Regno Lombardo–Veneto* (it)**
***Königreich Lombardo–Venetien* (de)**
***Österreichisches Italien* ("Austrian Italy")[]**

←  1815–1866  →

Flag

Coat of arms

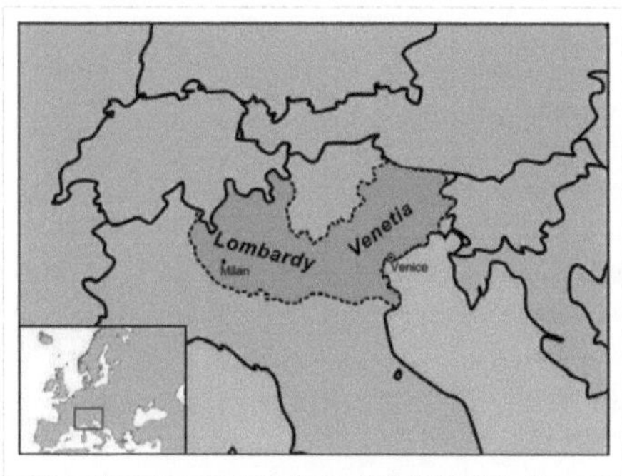

Outline of the Kingdom in 1852, superimposed on modern borders.

Capital		Milan and Venice (1815–1859) Venice (1859–1866)
Languages		Italian, German
Religion		Roman Catholic
Government		Absolute monarchy
King		
-	1815–1835	Francis I
-	1835–1848	Ferdinand I
-	1848–1866	Francis Joseph I
Viceroy		
-	1815	Heinrich XV
-	1857–1859	Ferdinand Joseph
History		

-	Congress of Vienna	9 June 1815
-	Five Days of Milan	22 March 1848
-	Treaty of Zürich	10 November 1859
-	Austro-Prussian War	14 June 1866
-	Peace of Prague	23 August 1866
-	Treaty of Vienna	12 October 1866
Area		
-	1852 []	46782 km² (18063 sq mi)
Population		
-	1852 est.[]	4671000
Density		99.8 /km² (258.6 /sq mi)
Currency		Lombardy-Venetia pound, (1816–1860) Lombardy-Venetia florin (1860–1866)
Today part of		Italy

The **Kingdom of Lombardy–Venetia** was created at the Congress of Vienna in recognition of the House of Habsburg-Lorraine's rights to Lombardy and the former Republic of Venice after the Napoleonic Kingdom of Italy, proclaimed in 1805, had collapsed.[] The kingdom was ruled day-to-day by viceroys appointed by the Imperial Court and resident in Milan and Venice.[] [] [] []

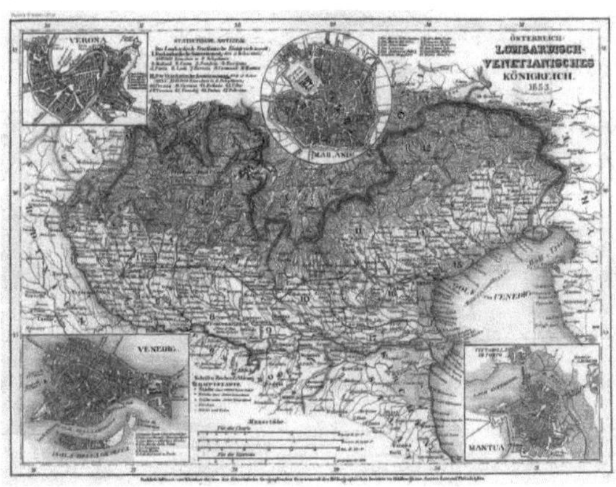

Lombardy and Venetia and the main cities of the Kingdom

History

The Congress of Vienna combined the territories of Lombardy (which had been ruled by the Habsburgs since the 16th century, and by the Austrian branch of the family from 1713 to 1796) and Venetia (which had been under Austrian rule intermittently since 1797) into a single unit under the House of Habsburg-Lorraine.

Administratively the Kingdom comprised two independent governments in the two parts. Lombardy included the provinces of Milan, Como, Bergamo, Brescia, Pavia, Cremona, Mantova, Lodi-Crema, and Sondrio. Venetia included the provinces of Venice, Verona, Padova, Vicenza, Treviso, Rovigo, Belluno, and Udine.[1]

The Kingdom of Lombardy–Venetia was first ruled by Francis I from 1815 to his death in 1835. Ferdinand I ruled from 1835 to 1848. In Milan on 6 September 1838 he became the last king to be crowned with the Iron Crown of Lombardy. The crown was subsequently brought to Vienna after the loss of Lombardy in 1859, but was restored to Italy after the loss of Venetia in 1866.

After a popular revolution on 22 March 1848 (*The Five Days of Milan*), the Austrians fled from Milan, which became the capital city of the *Governo Provvisorio della Lombardia* (Lombardy Provisional Government). The next day, Venice also rose against the Austrians, forming the *Governo Provvisorio di Venezia* (Venice Provisional Government). The Austrians, after defeating the Sardinian troops at the Battle of Custoza (24–25 July 1848), entered Milan (6 August) and Venice (24 August 1849), and restored Austrian rule.

Francis Joseph I ruled over the Kingdom for the rest of its existence. His younger brother Maximilian (who later became Emperor of Mexico), served as his viceroy in Milan between 1857 and 1859.

Lombardy was annexed to the embryonic Italian state in 1859, by the Treaty of Zurich after the Second Italian War of Independence; Venetia was ceded to the Kingdom of Italy in 1866 in the aftermath of the Seven Weeks War, by the Peace of Prague.[1]

Rulers

Viceroys

- Heinrich XV, Prince of Reuss-Plauen 1814–1815
- Count Friedrich Heinrich von Bellegarde 1815–1816
- Archduke Anton Victor of Austria 1816–1818
- Archduke Rainer of Austria 1818–1848
- Count Joseph Radetzky von Radetz 1848–1857
- Archduke Maximilian of Austria 1857–1859
- Ferencz Gyulai 1859

Governors of Lombardy

- Heinrich Johann Bellegarde 1814–1816
- Francesco Saurau 1816–1818
- Giulio Strassoldo di Sotto 1818–1830
- Franz Hartig 1830–1840
- Robert von Salm-Reifferscheidt-Raitz 1840–1841
- Johann Baptist Spaur 1841–1848
- Maximilian Karl Lamoral O'Donnell 1848 (acting)
- Felix von Schwarzenberg 1848
- Franz Wimpffen 1848 (acting)
- Alberto Montecuccoli-Laderchi 1848–1849 (acting)
- Karl Borromäus Philipp zu Schwarzenberg 1849–1850 (acting)
- Michele Strassoldo-Grafenberg 1851–1857 (with the title of Lieutenant of Lombardy)
- Friedrich von Burger 1857–1859

Governors of Venetia

- Peter Goëss 1815–1819
- Ferdinand Ernst Maria von Bissingen-Nippenburg 1819–1820
- Carlo d'Inzaghi 1820–1826
- Johann Baptist Spaur 1826–1840
- Aloys Pállfy de Erdöd 1840–1848
- Ferdinand Zichy zu Zich von Vasonykeöy 1848 (acting)
- Laval Nugent von Westmeath 1848–1849 (military governor)
- Karl von Gorzowsky 1849
- Stanislaus Anton Puchner 1849–1850
- Georg Otto von Toggenburg-Sargans 1850–1855
- Kajetan von Bissingen-Nippenburg 1855–1860
- Georg Otto von Toggenburg-Sargans 1860–1866 (second time)

Kings

Name	Lifespan	Reign start	Reign end	Notes	Family	Image
Francis I	12 February 1768 – 2 March 1835 (aged 67)	9 June 1815	2 March 1835		Habsburg-Lorraine	
Ferdinand I • *the Benign*	19 April 1793 – 29 June 1875 (aged 82)	2 March 1835	2 December 1848 (*abdicated*)	Son of Francis I	Habsburg-Lorraine	
Francis Joseph I	18 August 1830 – 21 November 1916 (aged 86)	2 December 1848	12 October 1866 (*deposed*)	Nephew of Ferdinand I	Habsburg-Lorraine	

Sources

[1] Rosita Rindler Schjerve (2003) "Diglossia and Power: Language Policies and Practice in the 19th Century Habsburg Empire", ISBN 3-11-017653-X, pp. 199-200 (http://books.google.com/books?id=w6W9V70JgvAC&pg=PA199&vq=Lombardy-Venetia&dq="utraquist+schools"&sig=RdPsUVamoozKVuLo5Er38FmKifw#PPA199,M1)

External links

- Flags of Lombardy–Venetia (http://www.worldstatesmen.org/Italy_states.htm#Lombardy)

Giubiasco

Giubiasco	

Country Switzerland Canton Ticino District Bellinzona	
Population	8364 (Dec 2011)[1]
- Density	1349 /km^2 (3494 /sq mi)
Area	km^2 (sq mi)[2]
Elevation	242 m (ft)
Postal code	6512
SFOS number	5005
Surrounded by	Bellinzona, Camorino, Gudo, Monte Carasso, Pianezzo, Sant'Antonino, Sementina
Twin towns	Prilly (Switzerland)
Website	www.giubiasco.ch [3] Profile [4] (Italian), SFSO statistics [5]
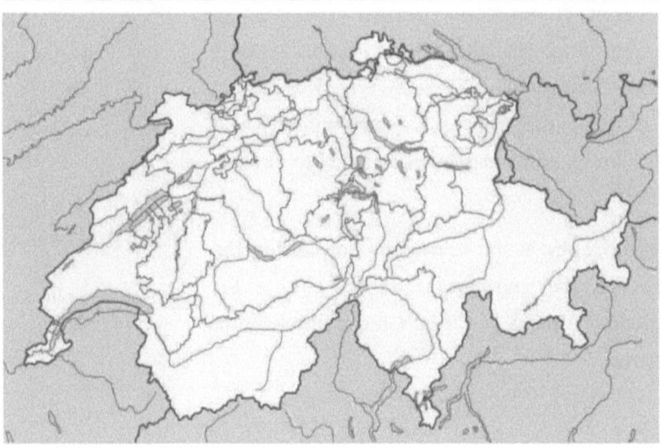 Giubiasco	

Giubiasco is a municipality in the district of Bellinzona in the canton of Ticino in Switzerland.

History

Due to its location at the center of numerous routes through the Alps, Giubiasco has been important since ancient times. The roads over the Monte Ceneri to the south, the Passo di San Jorio from the east, over the San Bernardino Pass and the Gotthard Pass from the north and the route along Lake Maggiore from the west all meet at Giubiasco.

Prehistoric necropolis

The discovery of the great cemetery of Giubiasco in 1900, led to uncontrolled treasure hunting and the loss of many objects. It was not until 1905 that the Swiss National Museum organized an excavation, which made it possible to save a certain number of graves and document the discoveries scientifically. Further, limited archaeological digs were organized in 1958 and 1971 to explore the edges of the graveyard. A large part of the collection is kept in the National Museum in Zurich, while other items have been given to various foreign and Swiss museums. The latest discoveries are in the Conservation Office in Bellinzona.[5]

The large necropolis, which was not far from the modern train station, has 565 graves. It is one of the rare examples of a site that was in continuous use for a particularly long period, stretching from the late Bronze Age (11th century BC) to the Roman period (2nd century AD). However, only the massive graveyard has been discovered, to date no associated settlement has been located. The cemetery has several different types of graves. The few Late Bronze Age graves are all cremation urn burials. The ashes of the deceased were buried in clay urns in simple trenches in the bare earth, the offerings are modest and consist almost exclusively of fragmented traditional ceramics.[5]

In the early Iron Age the Sopraceneri peoples burials gradually transformed from cremation to a full body burial. The graves were rectangular, bounded by dry stone trenches. From the mid-6th Century the distribution of grave goods reflect the Golasecca culture. They include a richer collection of items such as pottery (usually an urn, a bowl and a cup included) but only rarely any bronze containers. There were also clothing, accessories and jewelry; such as bronze or iron brooches, pendants, earrings, rings, belts, sheets and amber necklaces.

Most of the tombs in the necropolis date back to the Late Iron Age (4th-1st century BC) and are exclusively full body burials. From the beginning of the 3rd Century the influence of migration is visible in the graves. Celtic belt fittings, pendants and bronze bracelets are found in graves overlying ones from the Golasecca culture. Particularly interesting is the presence of weapons (spears and swords) and helmets of the early Iron Age, a period that is missing in other cemeteries (e.g. Solduno, which is now part of Locarno).

Even with the transition to Roman rule Giubiasco continued to play an important role, probably in connection with the military post of Castel Grande at Bellinzona. Roman items include imported bronze vessels and a silver bowl with two handles in a Roman style. Roman era graves included vessels of clay, metal or glass containing typical Roman grave goods. Additionally, coins and small iron day to day use items have been discovered.[5]

Medieval fortifications above the Roman fortress at Castel Grande

While the archeological discoveries at the necropolis end during the 2nd Century, it is likely that the graveyard remained in use for additional centuries. The nearby fortress of Castel Grande remained in use throughout the Roman era.

Medieval Giubiasco

During the Early Middle Ages, Giubiasco was probably the center of the land that the Abbey of San Pietro in Ciel d'Oro in Pavia, held in Locarno and the upper reaches of Ticino. Presumably, the church of St. Mary of Primasca, which is mentioned in a document of King Hugo of Italy from the 929 as the property of the monastery, was in the village.[6] The village of Giubiasco is first mentioned in 1186 as *apud Cibiascum*. In 1195 it was mentioned as *Zibiassco*.[7] By 1200, the Abbey had sold their ownership of the village and surrounding area to the noble Adam of Contone. At this time, the communities (*vicinanze*) of the valleys were able to partly buy limited self-rule with cash payments. In 1186, Frederick Barbarossa granted imperial immediacy to the region around Locarno.

After the battle of Arbedo in 1422, it fell under the power of Milan. From Milan, it was ruled by the Visconti and then by the Sforza families. Finally, after the Italian campaigns of the Swiss Confederation at the beginning of the 16th Century, it became part of the Bailiwick of Bellinzona, which was administratively shared between Uri, Schwyz, and Nidwalden.[6]

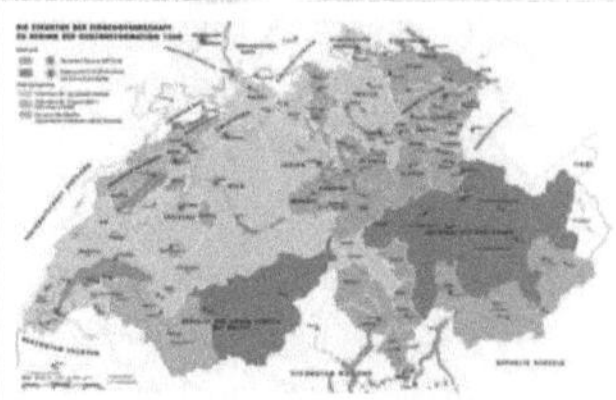

Swiss Confederation in 1560. Grey areas are "associated territories" which were administered by one or more of the full members of the Confederation

The Church of Santa Maria Assunta if first mentioned in 1387, but is probably from the 13th Century. In 1622 it was separated from Bellinzona and became a vice-parish church. It was raised in 1804 to become the parish church. Between the 15th and 17th Centuries, the church was enlarged and modified several times.

Early Modern Giubiasco

Originally, Giubiasco and the rest of the Valle Morobbia formed a single *vicinanze*, which split in the first half of the 16th Century. In 1831 divided the municipality of Morobbia Valle split into Valle Morobbia in Piano, Pianezzo and S. Antonio. Then, in 1867 Valle Morobbia in Piano merged into Giubiasco to form a single community.[6]

The economy of the municipality was based around raising crops and livestock, trading and some industry (grist mills and hammer mills). Some of the residents found jobs in other countries, particularly in Spain. Due to its location, Italian and German-Swiss livestock dealers, held livestock markets in Giubiasco starting in the first decades of the 16th Century. The municipality's privileges to hold the fair in the fall led to a protracted dispute with Lugano, which broke out in 1513 and was not settled until the end of the 19th Century.[6]

Modern Giubiasco

On 25 August 1814, the leaders of the liberal revolutionary people's movement in the Ticino met in Giubiasco, to resist the oligarchic cantonal constitution of 29 July. The so-called *Revolt of Giubiasco*, set up a temporary government but was soon crushed by Federal troops. The leaders were given heavy sentences. In 1853 the Camorino-Giubiasco-Sementina line was fortified, while the fortifications around Giubiasco were later dismantled, the *Fortini della fame* at Camorino is still visible. The 19th and 20th Century brought profound changes to the economy and society of the municipality. Due to its geographic location near Bellinzona, it became a station on the Gotthard railway. The connection to Locarno was completed in 1874, followed in 1882 by the Lugano and Luino route.[6]

Linoleum factory in Giubiasco

In 1905-06 two factories grew up next to the railway line; the linoleum works (later Forbo factory) and the Lenz engineering works (later Fischer e del San Gottardo Acciaierie Elettriche, closed 1925). In 1932, the Cattaneo AG ironworks opened. The ironworks still have a role in the regional economy.[6]

Agriculture has remained virtually unchanged, cereals, fruit and vegetable are grown of the plains and vineyards are on the hillside. A wine cooperative opened in 1928 and a market hall in 1937. Warehouses, mills and silos for an agricultural cooperative (Società cooperativa agricola ticinese, founded in 1941) supported other municipalities of the region.

The strong population growth in the 1970s and 80s, a consequence of the emergence of medium-sized industrial and small commercial businesses south of the main settlement, has slowed recently. The built-up area for housing and for commercial properties have extended from the old population centers at the foot of the mountain down into the valley floor. In 2000, about two-thirds of jobs were in the services sector, and about a fifth were in the manufacturing sector.[6]

Geography

Giubiasco has an area, as of 1997, of 6.23 square kilometers (sq mi). Of this area, 3.79 km^2 (sq mi) or 60.8% is used for agricultural purposes, while 1.32 km^2 (sq mi) or 21.2% is forested. Of the rest of the land, 2.23 km^2 (sq mi) or 35.8% is settled (buildings or roads), 0.09 km^2 (acres) or 1.4% is either rivers or lakes and 0.01 km^2 (acres) or 0.2% is unproductive land.

Giubiasco from the air

Of the built up area, industrial buildings made up 5.8% of the total area while housing and buildings made up 19.4% and transportation infrastructure made up 6.7%. Power and water infrastructure as well as other special developed areas made up 2.1% of the area while parks, green belts and sports fields made up 1.8%. Out of the forested land, 18.6% of the total land area is heavily forested and 2.6% is covered with orchards or small clusters of trees. Of the agricultural land, 36.0% is used for growing crops, while 5.5% is used for orchards or vine crops and 19.4% is used for alpine pastures. All the water in the municipality is flowing water.[8]

The municipality is located in the Bellinzona district, north of Bellinzona at the mouth of Valle Morobbia. It is part of the Agglomeration of Bellinzona. It consists of the villages of Giubiasco and Piano and the settlements of Lôro, Motti, Palasio, Pedevilla and Sasso Piatto. The municipality of Pianezzo is considering a merger some time in the future into Giubiasco.[9]

Coat of arms

The blazon of the municipal coat of arms is *Per pale Azure on a Base Vert beneath a Mountain of the same a house Argent roofed Gules and in Chief a Sun Or and Argent a Poplar eradicated proper.*[10]

Demographics

After Lugano, Bellinzona and Locarno, Giubiasco has one of the largest populations in Ticino.[7]

Giubiasco has a population (as of December 2011) of 8364.[1] As of 2008, 27.3% of the population are foreign nationals.[11] Over the last 10 years (1997–2007) the population has changed at a rate of 8.9%. Most of the population (as of 2000) speaks Italian(87.8%), with German being second most common (4.1%) and Serbo-Croatian being third (1.8%).[12]

Of the Swiss national languages (as of 2000), 302 speak German 72 people speak French, 6,516 people speak Italian, and 8 people speak Romansh. The remainder (520 people) speak another language.[13]

As of 2008, the gender distribution of the population was 47.7% male and 52.3% female. The population was made up of 2,816 Swiss men (34.0% of the population), and 1,134 (13.7%) non-Swiss men. There were 3,270 Swiss women (39.5%), and 1,065 (12.9%) non-Swiss women.[14]

In 2008 there were 60 live births to Swiss citizens and 19 births to non-Swiss citizens, and in same time span there were 69 deaths of Swiss citizens and 15 non-Swiss citizen deaths. Ignoring immigration and emigration, the population of Swiss citizens decreased by 9 while the foreign population increased by 4. There were 7 Swiss men who emigrated from Switzerland to another country, 9 Swiss women who emigrated from Switzerland to another country, 26 non-Swiss men who emigrated from Switzerland to another country and 15 non-Swiss women who emigrated from Switzerland to another country. The total Swiss population change in 2008 (from all sources) was an increase of 52 and the non-Swiss population change was an increase of 48 people. This represents a population growth rate of 1.2%.[11]

The age distribution, as of 2009, in Giubiasco is; 766 children or 9.2% of the population are between 0 and 9 years old and 888 teenagers or 10.7% are between 10 and 19. Of the adult population, 934 people or 11.3% of the population are between 20 and 29 years old. 1,187 people or 14.3% are between 30 and 39, 1,436 people or 17.3% are between 40 and 49, and 1,008 people or 12.2% are between 50 and 59. The senior population distribution is 921 people or 11.1% of the population are between 60 and 69 years old, 684 people or 8.3% are between 70 and 79, there are 461 people or 5.6% who are over 80.[14]

As of 2000, there were 3,273 private households in the municipality, and an average of 2.2 persons per household.[12] In 2000 there were 906 single family homes (or 59.1% of the total) out of a total of 1,534 inhabited buildings. There were 253 two family buildings (16.5%) and 251 multi-family buildings (16.4%). There were also 124 buildings in the municipality that were multipurpose buildings (used for both housing and commercial or another purpose).[15]

The vacancy rate for the municipality, in 2008, was 1.56%. In 2000 there were 3,628 apartments in the municipality. The most common apartment size was the 4 room apartment of which there were 1,222. There were 154 single room apartments and 623 apartments with five or more rooms.[16] Of these apartments, a total of 3,264 apartments (90.0% of the total) were permanently occupied, while 277 apartments (7.6%) were seasonally occupied and 87 apartments (2.4%) were empty.[16] As of 2007, the construction rate of new housing units was 10.4 new units per 1000 residents.[12]

The historical population is given in the following table:[7]

year	population
1591	ca. 1,000
1698	944
1801	950
1850	1,417
1900	1,722
1930	2,607
1950	3,311
1970	5,796
1990	6,897
1990	6,982
2000	7,418

Politics

In the 2007 federal election the most popular party was the FDP which received 32.36% of the vote. The next three most popular parties were the SP (22.06%), the CVP (18.55%) and the Ticino League (12.52%). In the federal election, a total of 2,162 votes were cast, and the voter turnout was 45.3%.[17]

In the 2007 *Gran Consiglio* election, there were a total of 4,653 registered voters in Giubiasco, of which 3,131 or 67.3% voted. 50 blank ballots and 3 null ballots were cast, leaving 3,078 valid ballots in the election. The most popular party was the PLRT which received 885 or 28.8% of the vote. The next three most popular parties were; the SSI (with 577 or 18.7%), the PS (with 550 or 17.9%) and the PPD+GenGiova (with 451 or 14.7%).[18]

In the 2007 *Consiglio di Stato* election, there were 34 blank ballots and 9 null ballots, which left 3,089 valid ballots in the election. The most popular party was the PLRT which received 805 or 26.1% of the vote. The next three most popular parties were; the PS (with 622 or 20.1%), the SSI (with 534 or 17.3%) and the PPD (with 492 or 15.9%).[18]

Economy

Train station at Giubiasco

As of 2007, Giubiasco had an unemployment rate of 5.23%. As of 2005, there were 78 people employed in the primary economic sector and about 20 businesses involved in this sector. 724 people are employed in the secondary sector and there are 75 businesses in this sector. 1,628 people are employed in the tertiary sector, with 277 businesses in this sector.[12]

There were 3,338 residents of the municipality who were employed in some capacity, of which females made up 42.0% of the workforce. In 2000, there were 1,941 workers who commuted into the municipality and 2,297 workers who commuted away. The municipality is a net exporter of workers, with about 1.2 workers leaving the municipality for every one entering. About 6.0% of the workforce coming into Giubiasco are coming from outside Switzerland, while 0.1% of the locals commute out of Switzerland for work.[19] Of the working population, 10.5% used public transportation to get to work, and 58.9% used a private car.[12]

As of 2009, there were 3 hotels in Giubiasco with a total of 26 rooms and 58 beds.[20]

Religion

From the 2000 census, 5,765 or 77.7% were Roman Catholic, while 323 or 4.4% belonged to the Swiss Reformed Church. There are 991 individuals (or about 13.36% of the population) who belong to another church (not listed on the census), and 339 individuals (or about 4.57% of the population) did not answer the question.[13]

Education

The entire Swiss population is generally well educated. In Giubiasco about 63% of the population (between age 25-64) have completed either non-mandatory upper secondary education or additional higher education (either university or a *Fachhochschule*).[12]

In Giubiasco there are a total of 1,404 students (as of 2009). The Ticino education system provides up to three years of non-mandatory kindergarten and in Giubiasco there are 209 children in kindergarten. The primary school program lasts for five years and includes both a standard school and a special school. In the municipality, 409 students attend the standard primary schools and 34 students attend the special school. In the lower secondary school system, students either attend a two-year middle school followed by a two-year pre-apprenticeship or they attend a four-year program to prepare for higher education. There are 347 students in the two-year middle school and 5 in their pre-apprenticeship, while 126 students are in the four-year advanced program.

The upper secondary school includes several options, but at the end of the upper secondary program, a student will be prepared to enter a trade or to continue on to a university or college. In Ticino, vocational students may either attend school while working on their internship or apprenticeship (which takes three or four years) or may attend school followed by an internship or apprenticeship (which takes one year as a full-time student or one and a half to two years as a part-time student).[21]

There are 84 vocational students who are attending school full-time and 169 who attend part-time.The professional program lasts three years and prepares a student for a job in engineering, nursing, computer science, business, tourism and similar fields. There are 21 students in the professional program.[22]

As of 2000, there were 476 students in Giubiasco who came from another municipality, while 343 residents attended schools outside the municipality.[19]

Transport

Giubiasco is served by the Giubiasco station, situated within the municipality. The station is on the Gotthard railway.

References

[1] Swiss Federal Statistics Office – STAT-TAB (http://www.pxweb.bfs.admin.ch/Dialog/varval.asp?ma=px-d-01-2A01&ti=Ständige+und+Nichtständige+Wohnbevölkerung+nach+Region,+Geschlecht,+Nationalität+und++Alter&path=../Database/German_01 - Bevölkerung/01.2 - Bevölkerungsstand und -bewegung/&lang=1&prod=01&openChild=true&secprod=2) *Ständige und Nichtständige Wohnbevölkerung nach Region, Geschlecht, Nationalität und Alter* (German) accessed 3 October 2012

[2] Arealstatistik Standard - Gemeindedaten nach 4 Hauptbereichen (http://www.bfs.admin.ch/bfs/portal/de/index/themen/02/03/blank/data/gemeindedaten.html)

[3] http://www.giubiasco.ch

[4] http://www.ti.ch/DFE/USTAT/DATI_COMUNI/0_pagine/default.asp?numero=5005

[5] *Giubiasco - Prehistoric and Roman eras* in German (http://www.hls-dhs-dss.ch/textes/d/D2034-1-1.php), French (http://www.hls-dhs-dss.ch/textes/f/F2034-1-1.php) and Italian (http://www.hls-dhs-dss.ch/textes/i/I2034-1-1.php) in the online *Historical Dictionary of Switzerland.*

[6] *Giubiasco - Medieval and Modern eras* in German (http://www.hls-dhs-dss.ch/textes/d/D2034-1-2.php), French (http://www.hls-dhs-dss.ch/textes/f/F2034-1-2.php) and Italian (http://www.hls-dhs-dss.ch/textes/i/I2034-1-2.php) in the online *Historical Dictionary of Switzerland.*

[7] *Giubiasco* in German (http://www.hls-dhs-dss.ch/textes/d/D2034.php), French (http://www.hls-dhs-dss.ch/textes/f/F2034.php) and Italian (http://www.hls-dhs-dss.ch/textes/i/I2034.php) in the online *Historical Dictionary of Switzerland.*

[8] Altitudine, superficie, secondo il genere di utilizzazione, rilevazione 1992/1997, e densità della popolazione, nel 2000 (http://www.ti.ch/DFE/USTAT/DATI_RIEPILOGHI/default_x.asp?menu=02&key1_testo=Territorio e ambiente&key1=02&key2=01&key3=01) (Italian) accessed 25 October 2010
[9] Amtliches Gemeindeverzeichnis der Schweiz (http://www.bfs.admin.ch/bfs/portal/de/index/infothek/nomenklaturen/blank/blank/gem_liste/03.html) published by the Swiss Federal Statistical Office (German) accessed 14 January 2010
[10] Flags of the World.com (http://www.fotw.us/flags/ch-ti005.html) accessed 26-October-2010
[11] Swiss Federal Statistical Office - Superweb database - Gemeinde Statistics 1981-2008 (http://www.bfs.admin.ch/bfs/portal/de/index/infothek/onlinedb/superweb/login.html) (German) accessed 19 June 2010
[12] Swiss Federal Statistical Office (http://www.bfs.admin.ch/bfs/portal/en/index/regionen/02/key.html) accessed 26-October-2010
[13] Popolazione residente, secondo la lingua principale e la religione, nel 2000 (http://www.ti.ch/DFE/USTAT/DATI_RIEPILOGHI/default_x.asp?menu=01&key1_testo=Popolazione&key1=01&key2=05&key3=00) (Italian) accessed 23 November 2010
[14] 01.02.03 Popolazione residente permanente (http://www.ti.ch/DFE/USTAT/DATI_RIEPILOGHI/default_x.asp?menu=01&key1_testo=Popolazione&key1=01&key2=02&key3=03) (Italian) accessed 23 November 2010
[15] 09.02.01 Edifici (http://www.ti.ch/DFE/USTAT/DATI_RIEPILOGHI/default_x.asp?menu=09&key1_testo=Costruzioni e abitazioni&key1=09&key2=02&key3=01) (Italian) accessed 23 November 2010
[16] 09.02.02 Abitazioni (http://www.ti.ch/DFE/USTAT/DATI_RIEPILOGHI/default_x.asp?menu=09&key1_testo=Costruzioni e abitazioni&key1=09&key2=02&key3=02) (Italian) accessed 23 November 2010
[17] Swiss Federal Statistical Office, *Nationalratswahlen 2007: Stärke der Parteien und Wahlbeteiligung, nach Gemeinden/Bezirk/Canton* (http://www.bfs.admin.ch/bfs/portal/de/index/themen/17/02/blank/data/04/03.html) (German) accessed 28 May 2010
[18] Elezioni cantonali: Gran Consiglio, Consiglio di Stato (http://www.ti.ch/DFE/USTAT/DATI_RIEPILOGHI/default_x.asp?menu=17&key1_testo=Politica&key1=17&key2=02&key3=02) (Italian) accessed 23 November 2010
[19] Swiss Federal Statistical Office - Statweb (http://www.media-stat.admin.ch/stat/pendler/pop.php) (German) accessed 24 June 2010
[20] Settori alberghiero e paralberghiero (http://www.ti.ch/DFE/USTAT/DATI_RIEPILOGHI/default_x.asp?menu=10&key1_testo=Turismo&key1=10&key2=03&key3=01) (Italian) accessed 23 November 2010
[21] EDK/CDIP/IDES (2010). KANTONALE SCHULSTRUKTUREN IN DER SCHWEIZ UND IM FÜRSTENTUM LIECHTENSTEIN / STRUCTURES SCOLAIRES CANTONALES EN SUISSE ET DANS LA PRINCIPAUTÉ DU LIECHTENSTEIN (http://edudoc.ch/record/35128/files/Schulsystem_alle.pdf) (Report).. Retrieved 24 June 2010.
[22] Allievi e studenti, secondo il genere di scuola, anno scolastico 2009/2010 (http://www.ti.ch/DFE/USTAT/DATI_RIEPILOGHI/default_x.asp?menu=15&key1_testo=Formazione e scienza&key1=15&key2=02&key3=01) (Italian) accessed 23 November 2010

External links

- Official website (http://www.giubiasco.ch) (Italian)
- *Giubiasco* in German (http://www.hls-dhs-dss.ch/textes/d/D2034.php), French (http://www.hls-dhs-dss.ch/textes/f/F2034.php) and Italian (http://www.hls-dhs-dss.ch/textes/i/I2034.php) in the online *Historical Dictionary of Switzerland*.

Cadenazzo

Cadenazzo	

Country Switzerland Canton Ticino District Bellinzona 	
Population	2401 (Dec 2011)[1]
- Density	287 /km^2 (743 /sq mi)
Area	km^2 (sq mi)[2]
Elevation	215 m (ft)
Postal code	6593
SFOS number	5003
Surrounded by	Bironico, Camorino, Contone, Cugnasco, Gudo, Isone, Locarno, Medeglia, Pianezzo, Ponte Capriasca, Rivera, Sant'Antonino, Sant'Antonio
Website	www.cadenazzo.ch [3] Profile [4] (Italian), SFSO statistics [5]

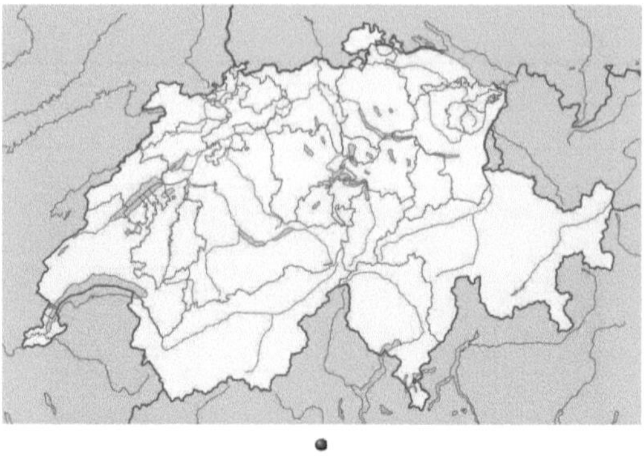

Cadenazzo

Cadenazzo is a municipality in the district of Bellinzona, in the canton of Ticino in Switzerland. On 13 March 2005 the formerly independent municipality of **Robasacco** was absorbed into Cadenazzo.

History

Cadenazzo is first mentioned in 1335 as *Catenacii*, when the *hospitallis de Cadenazio* was mentioned.[5] The existence of a medieval castle is in doubt. The collegiate church of San Pietro in Bellinzona owned part of the village by 1438. The chapel of S. Pietro Martire in Cadenazzo may have been built in the 13th Century, but the earliest mention of the chapel is from 1363. However, the chapel has not survived into the modern era and its location is unknown. In 1442, Cadenazzo separated from the mother church of Bellinzona joined the parish of the neighboring village of S. Antonino. They remained part of a joint parish until 1830.[5]

Agriculture, once dominant, but not very profitable, is still common in the Magadino valley. Crop yields increased after the correction of Ticino river. Since 1973, Cadenazzo has been home to a branch of the Swiss Federal Research Station for Agroecology and Agriculture. Due to some industrial enterprises, shops and warehouses in the vicinity of the train station, Cadenazzo draw commuters from surrounding communities.[5]

Geography

Cadenazzo has an area, as of 1997, of 8.37 square kilometers (sq mi). Of this area, 2.47 km^2 (sq mi) or 29.5% is used for agricultural purposes, while 2.05 km^2 (sq mi) or 24.5% is forested. Of the rest of the land, 1.25 km^2 (sq mi) or 14.9% is settled (buildings or roads), 0.11 km^2 (acres) or 1.3% is either rivers or lakes and 0.1 km^2 (acres) or 1.2% is unproductive land.

Of the built up area, industrial buildings made up 3.7% of the total area while housing and buildings made up 4.9% and transportation infrastructure made up 5.1%. Out of the forested land, 22.8% of the total land area is heavily forested and 1.1% is covered with orchards or small clusters of trees. Of the agricultural land, 20.8% is used for growing crops, while 3.8% is used for orchards or vine crops and 4.9% is used for alpine pastures. All the water in the municipality is flowing water. Of the unproductive areas, 1.1% is unproductive vegetation.[6]

The municipality is located in the Bellinzona district, along a road along the north foot of the Monte Ceneri.

Coat of arms

The blazon of the municipal coat of arms is *Azure a Pale Argent and overall a Castle embatteled Gules.*[7]

Demographics

Cadenazzo has a population (as of December 2011) of 2401.[1] As of 2008, 36.2% of the population are foreign nationals.[8] Over the last 10 years (1997–2007) the population has changed at a rate of 18.9%. Most of the population (as of 2000) speaks Italian(81.4%), with German being second most common (5.2%) and Albanian being third (3.3%).[9]

Of the Swiss national languages (as of 2000), 92 speak German 27 people speak French, 1,417 people speak Italian, and 1 person speaks Romansh. The remainder (218 people) speak another language.[10]

As of 2008, the gender distribution of the population was 51.2% male and 48.8% female. The population was made up of 749 Swiss men (31.9% of the population), and 453 (19.3%) non-Swiss men. There were 759 Swiss women (32.3%), and 386 (16.4%) non-Swiss women.[11]

In 2008 there were 15 live births to Swiss citizens and 5 births to non-Swiss citizens, and in same time span there were 10 deaths of Swiss citizens and 3 non-Swiss citizen deaths. Ignoring immigration and emigration, the population of Swiss citizens increased by 5 while the foreign population increased by 2. There were 2 Swiss men who emigrated from Switzerland to another country, 2 Swiss women who emigrated from Switzerland to another country, 13 non-Swiss men who emigrated from Switzerland to another country and 12 non-Swiss women who emigrated from Switzerland to another country. The total Swiss population change in 2008 (from all sources) was an increase of 76 and the non-Swiss population change was a decrease of 3 people. This represents a population growth rate of 3.3%.[8]

The age distribution, as of 2009, in Cadenazzo is; 256 children or 10.9% of the population are between 0 and 9 years old and 255 teenagers or 10.9% are between 10 and 19. Of the adult population, 333 people or 14.2% of the population are between 20 and 29 years old. 330 people or 14.1% are between 30 and 39, 427 people or 18.2% are between 40 and 49, and 345 people or 14.7% are between 50 and 59. The senior population distribution is 196 people or 8.4% of the population are between 60 and 69 years old, 115 people or 4.9% are between 70 and 79, there are 90 people or 3.8% who are over 80.[11]

As of 2000, there were 741 private households in the municipality, and an average of 2.4 persons per household.[9] In 2000 there were 274 single family homes (or 67.7% of the total) out of a total of 405 inhabited buildings. There were 55 two family buildings (13.6%) and 41 multi-family buildings (10.1%). There were also 35 buildings in the municipality that were multipurpose buildings (used for both housing and commercial or another purpose).[12]

The vacancy rate for the municipality, in 2008, was 1.33%. In 2000 there were 805 apartments in the municipality. Of these apartments, a total of 683 apartments (84.8% of the total) were permanently occupied, while 99 apartments (12.3%) were seasonally occupied and 23 apartments (2.9%) were empty.[13] The most common apartment size was the 4 room apartment of which there were 293. There were 27 single room apartments and 134 apartments with five or more rooms.[13] As of 2007, the construction rate of new housing units was 28.6 new units per 1000 residents.[9]

The historical population is given in the following table:[5]

year	population
1698	191
1784	200
1850	216
1900	333
1950	621
1990	1,500
2000	1,755

Politics

In the 2007 federal election the most popular party was the FDP which received 38.74% of the vote. The next three most popular parties were the CVP (23.01%), the Ticino League (13.22%) and the SP (11.64%). In the federal election, a total of 479 votes were cast, and the voter turnout was 42.3%.[14]

In the 2007 *Gran Consiglio* election, there were a total of 1,113 registered voters in Cadenazzo, of which 731 or 65.7% voted. 11 blank ballots and 1 null ballots were cast, leaving 719 valid ballots in the election. The most popular party was the PLRT which received 252 or 35.0% of the vote. The next three most popular parties were; the PPD+GenGiova (with 124 or 17.2%), the SSI (with 118 or 16.4%) and the LEGA (with 94 or 13.1%).[15]

In the 2007 *Consiglio di Stato* election, there were 5 blank ballots and 2 null ballots, which left 724 valid ballots in the election. The most popular party was the PLRT which received 236 or 32.6% of the vote. The next three most popular parties were; the LEGA (with 152 or 21.0%), the PPD (with 130 or 18.0%) and the SSI (with 88 or 12.2%).[15]

Economy

As of 2007, Cadenazzo had an unemployment rate of 5.58%. As of 2005, there were 83 people employed in the primary economic sector and about 20 businesses involved in this sector. 181 people are employed in the secondary sector and there are 27 businesses in this sector. 634 people are employed in the tertiary sector, with 112 businesses in this sector.[9]

There were 897 residents of the municipality who were employed in some capacity, of which females made up 36.8% of the workforce. In 2000, there were 917 workers who commuted into the municipality and 600 workers who commuted away. The municipality is a net importer of workers, with about 1.5 workers entering the municipality for every one leaving. About 13.0% of the workforce coming into Cadenazzo are coming from outside Switzerland.[16] Of the working population, 8.4% used public transportation to get to work, and 52.7% used a private car.[9]

As of 2009, there was one hotel in Cadenazzo.[17]

Religion

From the 2000 census, 1,298 or 74.0% were Roman Catholic, while 95 or 5.4% belonged to the Swiss Reformed Church. There are 302 individuals (or about 17.21% of the population) who belong to another church (not listed on the census), and 60 individuals (or about 3.42% of the population) did not answer the question.[10]

Climate

Between 1961 and 1990 Cadenazzo had an average of 103 days of rain or snow per year and on average received 1772 mm (in) of precipitation. The wettest month was May during which time Magadino received an average of 211 mm (in) of rain or snow. During this month there was precipitation for an average of 12.9 days. The driest month of the year was December with an average of 67 mm (in) of precipitation over 5.5 days.[18]

Education

In Cadenazzo about 55.2% of the population (between age 25-64) have completed either non-mandatory upper secondary education or additional higher education (either University or a *Fachhochschule*).[9]

In Cadenazzo there are a total of 440 students (as of 2009). The Ticino education system provides up to three years of non-mandatory kindergarten and in Cadenazzo there are 62 children in kindergarten.

The primary school program lasts for five years and includes both a standard school and a special school. In the municipality, 137 students attend the standard primary schools and 11 students attend the special school. In the lower secondary school system, students either attend a two-year middle school followed by a two-year pre-apprenticeship or they attend a four-year program to prepare for higher education. There are 119 students in the two-year middle school and 1 in their pre-apprenticeship, while 19 students are in the four-year advanced program.

The upper secondary school includes several options, but at the end of the upper secondary program, a student will be prepared to enter a trade or to continue on to a university or college. In Ticino, vocational students may either attend school while working on their internship or apprenticeship (which takes three or four years) or may attend school followed by an internship or apprenticeship (which takes one year as a full-time student or one and a half to two years as a part-time student).[20]

There are 22 vocational students who are attending school full-time and 67 who attend part-time.The professional program lasts three years and prepares a student for a job in engineering, nursing, computer science, business, tourism and similar fields. There are 2 students in the professional program.[21]

As of 2000, there were 308 students in Cadenazzo who came from another municipality, while 67 residents attended schools outside the municipality.[16]

References

[1] Swiss Federal Statistics Office – STAT-TAB (http://www.pxweb.bfs.admin.ch/Dialog/varval.asp?ma=px-d-01-2A01&ti=Ständige+und+Nichtständige+Wohnbevölkerung+nach+Region,+Geschlecht,+Nationalität+und++Alter&path=../Database/German_01 - Bevölkerung/01.2 - Bevölkerungsstand und -bewegung/&lang=1&prod=01&openChild=true&secprod=2) *Ständige und Nichtständige Wohnbevölkerung nach Region, Geschlecht, Nationalität und Alter* (German) accessed 3 October 2012

[2] Arealstatistik Standard - Gemeindedaten nach 4 Hauptbereichen (http://www.bfs.admin.ch/bfs/portal/de/index/themen/02/03/blank/data/gemeindedaten.html)

[3] http://www.cadenazzo.ch

[4] http://www.ti.ch/DFE/USTAT/DATI_COMUNI/0_pagine/default.asp?numero=5003

[5] *Cadenazzo* in German (http://www.hls-dhs-dss.ch/textes/d/D2032.php), French (http://www.hls-dhs-dss.ch/textes/f/F2032.php) and Italian (http://www.hls-dhs-dss.ch/textes/i/I2032.php) in the online *Historical Dictionary of Switzerland.*

[6] Altitudine, superficie, secondo il genere di utilizzazione, rilevazione 1992/1997, e densità della popolazione, nel 2000 (http://www.ti.ch/DFE/USTAT/DATI_RIEPILOGHI/default_x.asp?menu=02&key1_testo=Territorio e ambiente&key1=02&key2=01&key3=01) (Italian) accessed 25 October 2010

[7] Flags of the World.com (http://www.fotw.us/flags/ch-ti003.html) accessed 26-October-2010

[8] Swiss Federal Statistical Office - Superweb database - Gemeinde Statistics 1981-2008 (http://www.bfs.admin.ch/bfs/portal/de/index/infothek/onlinedb/superweb/login.html) (German) accessed 19 June 2010
[9] Swiss Federal Statistical Office (http://www.bfs.admin.ch/bfs/portal/en/index/regionen/02/key.html) accessed 26-October-2010
[10] Popolazione residente, secondo la lingua principale e la religione, nel 2000 (http://www.ti.ch/DFE/USTAT/DATI_RIEPILOGHI/default_x.asp?menu=01&key1_testo=Popolazione&key1=01&key2=05&key3=00) (Italian) accessed 23 November 2010
[11] 01.02.03 Popolazione residente permanente (http://www.ti.ch/DFE/USTAT/DATI_RIEPILOGHI/default_x.asp?menu=01&key1_testo=Popolazione&key1=01&key2=02&key3=03) (Italian) accessed 23 November 2010
[12] 09.02.01 Edifici (http://www.ti.ch/DFE/USTAT/DATI_RIEPILOGHI/default_x.asp?menu=09&key1_testo=Costruzioni e abitazioni&key1=09&key2=02&key3=01) (Italian) accessed 23 November 2010
[13] 09.02.02 Abitazioni (http://www.ti.ch/DFE/USTAT/DATI_RIEPILOGHI/default_x.asp?menu=09&key1_testo=Costruzioni e abitazioni&key1=09&key2=02&key3=02) (Italian) accessed 23 November 2010
[14] Swiss Federal Statistical Office, *Nationalratswahlen 2007: Stärke der Parteien und Wahlbeteiligung, nach Gemeinden/Bezirk/Canton* (http://www.bfs.admin.ch/bfs/portal/de/index/themen/17/02/blank/data/04/03.html) (German) accessed 28 May 2010
[15] Elezioni cantonali: Gran Consiglio, Consiglio di Stato (http://www.ti.ch/DFE/USTAT/DATI_RIEPILOGHI/default_x.asp?menu=17&key1_testo=Politica&key1=17&key2=02&key3=02) (Italian) accessed 23 November 2010
[16] Swiss Federal Statistical Office - Statweb (http://www.media-stat.admin.ch/stat/pendler/pop.php) (German) accessed 24 June 2010
[17] Settori alberghiero e paralberghiero (http://www.ti.ch/DFE/USTAT/DATI_RIEPILOGHI/default_x.asp?menu=10&key1_testo=Turismo&key1=10&key2=03&key3=01) (Italian) accessed 23 November 2010
[18] .
[19] The weather station elevation is 203 meters above sea level.
[20] EDK/CDIP/IDES (2010). KANTONALE SCHULSTRUKTUREN IN DER SCHWEIZ UND IM FÜRSTENTUM LIECHTENSTEIN / STRUCTURES SCOLAIRES CANTONALES EN SUISSE ET DANS LA PRINCIPAUTÉ DU LIECHTENSTEIN (http://edudoc.ch/record/35128/files/Schulsystem_alle.pdf) (Report). . Retrieved 24 June 2010.
[21] Allievi e studenti, secondo il genere di scuola, anno scolastico 2009/2010 (http://www.ti.ch/DFE/USTAT/DATI_RIEPILOGHI/default_x.asp?menu=15&key1_testo=Formazione e scienza&key1=15&key2=02&key3=01) (Italian) accessed 23 November 2010

External links

- Official website (http://www.cadenazzo.ch) (Italian)
- *Cadenazzo* in German (http://www.hls-dhs-dss.ch/textes/d/D2032.php), French (http://www.hls-dhs-dss.ch/textes/f/F2032.php) and Italian (http://www.hls-dhs-dss.ch/textes/i/I2032.php) in the online *Historical Dictionary of Switzerland*.

Article Sources and Contributors

Camorino *Source*: http://en.wikipedia.org/w/index.php?title=Camorino *Contributors*: Arjayay, Arota, BillC, D6, Khazar2, Ksnow, Tobyc75, 1 anonymous edits

Cantons_of_Switzerland *Source*: http://en.wikipedia.org/w/index.php?title=Cantons_of_Switzerland *Contributors*: 05, 159753, 2Neuenburg, 3riol, A bit iffy, A. Parrot, AP1787, Acntx, Ahuskay, Alarob, Aram33, Argasp, Arthena, Bag, Barticus88, Bearcat, Blanche of King's Lynn, Brammen, Briséis, Brittainia, BrownHairedGirl, CUSENZA Mario, Cassowary, Chanheigeorge, Chris j wood, ChrisfromHouston, Chùn-hiàn, Claude girardin, Closedmouth, Cnilep, Cunado19, D6, DDima, Dbachmann, Dcoetzee, Deacon of Pndapetzim, Denelson83, Djinn112, Docu, Dzordzm, EamonnPKeane, EdwinHJ, El C, Electionworld, Enlil Ninlil, Ericamick, Fan-1967, Filzstift, Frank101, Gazilion, Gestumblindi, Gurch, Haddiscoe, Hashar, Hayden120, Hinote, Hmains, Hroðulf, Hunnjazal, Hvn0413, Indigoeleven, Insilvis, Janburse, JdeJ, Jschwa1, Jusjih, Kahusi, Kerryh r, Kokiri, Korovioff, Lightmouse, Lkmorlan, Longbow4u, Luuva, MZMcBride, MadGeographer, Martg76, Mascia Luccova, Master of Puppets, Math Champion, Maxí, Megan1967, MeltBanana, Mistakefinder, MrRadioGuy, Mrt3366, Mschlindwein, Mthwaite, Murtasa, N2e, Nalta, Ngagnebin, Nilfanion, Nohat, Nowhither, Ojw, OwenBlacker, Pascal, Patrick, Paul-L, Piccadilly, Piccolo Modificatore Laborioso, Pne, Postdlf, PsY.cHo, Random user 39849958, Rich Farmbrough, Rjwilmsi, Roke, Rpckvv, Sallygirl1783, Sam Blacketer, Sandstein, Sardanaphalus, Schutz, ScottMainwaring, Sesel, SetRochele, SetRochelewsad, Sherool, SiBr4, Skew-t, Sluzzelin, Smjg, Stephensuleeman, Stevietheman, Tamfang, TenOfAllTrades, Thorsen, Tobias Conradi, Tobyc75, Tonymec, Ultimate Destiny, Uncle Scrooge, Vdiest, West.andrew.g, Woohookitty, Worobiew, Wshun, Zacharie Grossen, Zondor, Žiga, ვაგა მ, 104 anonymous edits

Bellinzona_(district) *Source*: http://en.wikipedia.org/w/index.php?title=Bellinzona_%28district%29 *Contributors*: 2Neuenburg, Cate, Chris j wood, Diamond2, DidiWeidmann, Khazar2, Knepflerle, Saltose, Schwyz, Tobyc75, Ultimate Destiny, 6 anonymous edits

Communes_of_Switzerland *Source*: http://en.wikipedia.org/w/index.php?title=Communes_of_Switzerland *Contributors*: 159753, 2Neuenburg, Acntx, Amikake3, Andonic, Booksworm, Chanheigeorge, Corti, Crusoe8181, Cusio, D6, Db1944, Dbachmann, Dirrival, Docu, Elipongo, Emerson7, EnOreg, Felix Portier, Frietjes, Geschichte, Gestumblindi, Grutness, Hadal, Haruo, Irmgard, J. 'mach' wust, Jurema Oliveira, LeoNomis, Lugia2453, MBisanz, MK, Megan1967, Melo86, NawlinWiki, Onco p53, PsY.cHo, Reindra, Rich Farmbrough, Roland zh, RubleuleR, Schutz, Sectryan, Sietse Snel, Srleffler, Tobias Conradi, Tobyc75, Ultimate Destiny, Underlying lk, Wmahan, Woohookitty, Yellowdesk, 27 anonymous edits

Sementina *Source*: http://en.wikipedia.org/w/index.php?title=Sementina *Contributors*: BillC, Cate, D6, Dl2000, Docu, Khazar2, Ksnow, Rich Farmbrough, Sole Soul, Tobyc75, Zanmatt, 2 anonymous edits

Bellinzona *Source*: http://en.wikipedia.org/w/index.php?title=Bellinzona *Contributors*: Ale jrb, Andrwsc, Attilios, Bahnfrend, BillC, Cerb, Chris j wood, ChrisGualtieri, CommonsDelinker, Corti, Cs-wolves, D6, DF08, Darius Dhlomo, Dbachmann, Diwas, Docu, Euratlas, FS Italia, Firsfron, Futbol vic, Gargolla, Ghirlandajo, GoingBatty, Goldfishbutt, Hmendonca, Italian boy, Jauhienij, JdeJ, Jeppiz, Joey80, JustAGal, Keke Ruehle, Khazar2, Kiranchikkadi, Ksnow, Kummi, Kwamikagami, Lotje, Lupin, MadGeographer, Marco Bernasocchi, Marek69, Matthew hk, Mdw0, MrBoo, Nick Number, Nova77, Ohconfucius, Ori, OwenBlacker, Pasquale, Patch051, Piccolo Modificatore Laborioso, Rhymezz, Rich Farmbrough, Rjwilmsi, Sandstein, SchreiberBike, Sun Creator, Timc, Tobyc75, Trilobite, Welsh, Widr, Zacharie Grossen, Zeno Crivelli, Zipacna1, Прон, 48 anonymous edits

Kingdom_of_Lombardy–Venetia *Source*: http://en.wikipedia.org/w/index.php?title=Kingdom_of_Lombardy%E2%80%93Venetia *Contributors*: Aamsse, AjaxSmack, Altenmann, ArnoldPlaton, Attilios, Caerwine, Chochopk, ChryZ MUC, Cladeal832, Cusio, Da Joe, Daufer, DeCausa, Domino theory, Dove1950, Fawcett5, Filippusson, Frokor, Garion96, Gennarous, Gentaur, Ghepeu, Giovanni Giove, Greenshed, Gryffindor, Gugganij, Hans Adler, Jacquesverlaeken, Japanese Searobin, Jfruh, Jllm06, Kaiser Torikka, LoveActresses, LtNOWIS, Lukeno94, Luwilt, M.O.X, Mattis, MeltBanana, Mikeygolf015, Mrlopez2681, Mvaldemar, Neddyseagoon, OwenBlacker, P. S. Burton, Panairjdde, Pasquale, Pearle, Qbli2mHd, R-41, Regnator, Rich Farmbrough, SchreiberBike, Slb nsk, Srnec, TRAJAN 117, Volker89, Wetman, Worobiew, 27 anonymous edits

Giubiasco *Source*: http://en.wikipedia.org/w/index.php?title=Giubiasco *Contributors*: BillC, Cate, Chris j wood, D6, Deny2010, Docu, Hughesdavidw, Khazar2, Ksnow, Marek69, Nick Number, Tillo, Tobyc75, 2 anonymous edits

Cadenazzo *Source*: http://en.wikipedia.org/w/index.php?title=Cadenazzo *Contributors*: BillC, D6, FS Italia, Khazar2, Ksnow, Rich Farmbrough, Roland zh, Stephenchou0722, Tobyc75, 3 anonymous edits

Image Sources, Licenses and Contributors

file:Camorino-coat of arms.svg *Source*: http://en.wikipedia.org/w/index.php?title=File:Camorino-coat_of_arms.svg *License*: unknown *Contributors*: User:Delta-9

file:Switzerland location map.svg *Source*: http://en.wikipedia.org/w/index.php?title=File:Switzerland_location_map.svg *License*: unknown *Contributors*: Lukasb1992, Maoman, NordNordWest, Sting, Wikisoft*, 1 anonymous edits

File:Red pog.svg *Source*: http://en.wikipedia.org/w/index.php?title=File:Red_pog.svg *License*: unknown *Contributors*: Anomie

File:Ecke in Camorino TI 1932.jpg *Source*: http://en.wikipedia.org/w/index.php?title=File:Ecke_in_Camorino_TI_1932.jpg *License*: unknown *Contributors*: Peter Forster

File:Camorino TI 1932 Svizzera, Switzerland.jpg *Source*: http://en.wikipedia.org/w/index.php?title=File:Camorino_TI_1932_Svizzera,_Switzerland.jpg *License*: unknown *Contributors*: Peter Forster

File:Suisse cantons.svg *Source*: http://en.wikipedia.org/w/index.php?title=File:Suisse_cantons.svg *License*: unknown *Contributors*: User:Pymouss44

File:Karte 13 Alte Orte.png *Source*: http://en.wikipedia.org/w/index.php?title=File:Karte_13_Alte_Orte.png *License*: unknown *Contributors*: User:Sidonius

Image:Wappen Zürich matt.svg *Source*: http://en.wikipedia.org/w/index.php?title=File:Wappen_Zürich_matt.svg *License*: unknown *Contributors*: Sa-se

Image:Wappen Bern matt.svg *Source*: http://en.wikipedia.org/w/index.php?title=File:Wappen_Bern_matt.svg *License*: unknown *Contributors*: User:Sa-se

Image:Wappen Luzern matt.svg *Source*: http://en.wikipedia.org/w/index.php?title=File:Wappen_Luzern_matt.svg *License*: unknown *Contributors*: Sa-se

Image:Wappen Uri matt.svg *Source*: http://en.wikipedia.org/w/index.php?title=File:Wappen_Uri_matt.svg *License*: unknown *Contributors*: Sa-se

Image:Wappen Obwalden matt.svg *Source*: http://en.wikipedia.org/w/index.php?title=File:Wappen_Obwalden_matt.svg *License*: unknown *Contributors*: Heraldiker Rudolf Münger

Image:Wappen Nidwalden matt.svg *Source*: http://en.wikipedia.org/w/index.php?title=File:Wappen_Nidwalden_matt.svg *License*: unknown *Contributors*: Sa-se

Image:Wappen Glarus matt.svg *Source*: http://en.wikipedia.org/w/index.php?title=File:Wappen_Glarus_matt.svg *License*: unknown *Contributors*: Sa-se

Image:Wappen Zug matt.svg *Source*: http://en.wikipedia.org/w/index.php?title=File:Wappen_Zug_matt.svg *License*: unknown *Contributors*: Sa-se

Image:Wappen Freiburg matt.svg *Source*: http://en.wikipedia.org/w/index.php?title=File:Wappen_Freiburg_matt.svg *License*: unknown *Contributors*: Sa-se

Image:Wappen Solothurn matt.svg *Source*: http://en.wikipedia.org/w/index.php?title=File:Wappen_Solothurn_matt.svg *License*: unknown *Contributors*: Sa-se

Image:Wappen Basel-Stadt matt.svg *Source*: http://en.wikipedia.org/w/index.php?title=File:Wappen_Basel-Stadt_matt.svg *License*: unknown *Contributors*: Sa-se

Image:Coat of arms of Kanton Basel-Landschaft.svg *Source*: http://en.wikipedia.org/w/index.php?title=File:Coat_of_arms_of_Kanton_Basel-Landschaft.svg *License*: unknown *Contributors*: Wappenentwurf: Otto Plattner diese Datei: Jwnabd

Image:Wappen Schaffhausen matt.svg *Source*: http://en.wikipedia.org/w/index.php?title=File:Wappen_Schaffhausen_matt.svg *License*: unknown *Contributors*: (previous version maybe User:Delta-9)

Image:Wappen Appenzell Ausserrhoden matt.svg *Source*: http://en.wikipedia.org/w/index.php?title=File:Wappen_Appenzell_Ausserrhoden_matt.svg *License*: unknown *Contributors*: Sa-se

Image:Wappen Appenzell Innerrhoden matt.svg *Source*: http://en.wikipedia.org/w/index.php?title=File:Wappen_Appenzell_Innerrhoden_matt.svg *License*: unknown *Contributors*: Sa-se

Image:Coat of arms of canton of St. Gallen.svg *Source*: http://en.wikipedia.org/w/index.php?title=File:Coat_of_arms_of_canton_of_St._Gallen.svg *License*: unknown *Contributors*: Wappenentwurf: unbekannt diese Datei: Jwnabd

Image:Wappen Graubünden matt.svg *Source*: http://en.wikipedia.org/w/index.php?title=File:Wappen_Graubünden_matt.svg *License*: unknown *Contributors*: Sa-se

Image:Wappen Aargau matt.svg *Source*: http://en.wikipedia.org/w/index.php?title=File:Wappen_Aargau_matt.svg *License*: unknown *Contributors*: Sa-se

Image:Wappen Thurgau matt.svg *Source*: http://en.wikipedia.org/w/index.php?title=File:Wappen_Thurgau_matt.svg *License*: unknown *Contributors*: Sa-se

Image:Wappen Tessin matt.svg *Source*: http://en.wikipedia.org/w/index.php?title=File:Wappen_Tessin_matt.svg *License*: unknown *Contributors*: Abaddon1337, Ricordisamoa, Sa-se, Skim, 1 anonymous edits

Image:Wappen Waadt matt.svg *Source*: http://en.wikipedia.org/w/index.php?title=File:Wappen_Waadt_matt.svg *License*: unknown *Contributors*: Sa-se

Image:Wappen Wallis matt.svg *Source*: http://en.wikipedia.org/w/index.php?title=File:Wappen_Wallis_matt.svg *License*: unknown *Contributors*: Sa-se

Image:Wappen Neuenburg matt.svg *Source*: http://en.wikipedia.org/w/index.php?title=File:Wappen_Neuenburg_matt.svg *License*: unknown *Contributors*: Sa-se

Image:Wappen Genf matt.svg *Source*: http://en.wikipedia.org/w/index.php?title=File:Wappen_Genf_matt.svg *License*: unknown *Contributors*: Sa-se

Image:Wappen Jura matt.svg *Source*: http://en.wikipedia.org/w/index.php?title=File:Wappen_Jura_matt.svg *License*: unknown *Contributors*: Sa-se

Image:Coat of Arms of Switzerland.svg *Source*: http://en.wikipedia.org/w/index.php?title=File:Coat_of_Arms_of_Switzerland.svg *License*: unknown *Contributors*: User:E Pluribus Anthony, User:Reisio, User:Wiki-vr, User:Zscout370

Image:Karikatur Teilung Basels.jpg *Source*: http://en.wikipedia.org/w/index.php?title=File:Karikatur_Teilung_Basels.jpg *License*: unknown *Contributors*: Alarob, Aliman5040, Infrogmation, Mapmarks, Mattes, Parpan05, Roland zh, Sidonius, Stefan Kühn

File:Snake-coat of arms.svg *Source*: http://en.wikipedia.org/w/index.php?title=File:Snake-coat_of_arms.svg *License*: unknown *Contributors*: User:Delta-9

File:ArbedoCastione-coat of arms.svg *Source*: http://en.wikipedia.org/w/index.php?title=File:ArbedoCastione-coat_of_arms.svg *License*: unknown *Contributors*: User:Delta-9

File:Lumino-coat of arms.svg *Source*: http://en.wikipedia.org/w/index.php?title=File:Lumino-coat_of_arms.svg *License*: unknown *Contributors*: User:Delta-9

File:MonteCarasso-coat of arms.svg *Source*: http://en.wikipedia.org/w/index.php?title=File:MonteCarasso-coat_of_arms.svg *License*: unknown *Contributors*: User:Delta-9

File:Gudo-coat of arms.svg *Source*: http://en.wikipedia.org/w/index.php?title=File:Gudo-coat_of_arms.svg *License*: unknown *Contributors*: User:Delta-9

File:Sementina-coat of arms.svg *Source*: http://en.wikipedia.org/w/index.php?title=File:Sementina-coat_of_arms.svg *License*: unknown *Contributors*: User:Delta-9

File:Gorduno-coat of arms.svg *Source*: http://en.wikipedia.org/w/index.php?title=File:Gorduno-coat_of_arms.svg *License*: unknown *Contributors*: User:Delta-9

File:Gnosca-coat of arms.svg *Source*: http://en.wikipedia.org/w/index.php?title=File:Gnosca-coat_of_arms.svg *License*: unknown *Contributors*: User:Delta-9

File:Preonzo-coat of arms.svg *Source*: http://en.wikipedia.org/w/index.php?title=File:Preonzo-coat_of_arms.svg *License*: unknown *Contributors*: User:Delta-9

File:Moleno-coat of arms.svg *Source*: http://en.wikipedia.org/w/index.php?title=File:Moleno-coat_of_arms.svg *License*: unknown *Contributors*: User:Delta-9

File:Giubiasco-coat of arms.svg *Source*: http://en.wikipedia.org/w/index.php?title=File:Giubiasco-coat_of_arms.svg *License*: unknown *Contributors*: User:Delta-9

File:Pianezzo-coat of arms.svg *Source*: http://en.wikipedia.org/w/index.php?title=File:Pianezzo-coat_of_arms.svg *License*: unknown *Contributors*: User:Delta-9

File:SantAntonio-coat of arms.svg *Source*: http://en.wikipedia.org/w/index.php?title=File:SantAntonio-coat_of_arms.svg *License*: unknown *Contributors*: User:Delta-9

File:Camorino-coat of arms.svg *Source*: http://en.wikipedia.org/w/index.php?title=File:Camorino-coat_of_arms.svg *License*: unknown *Contributors*: User:Delta-9

File:SantAntonino-coat of arms.svg *Source*: http://en.wikipedia.org/w/index.php?title=File:SantAntonino-coat_of_arms.svg *License*: unknown *Contributors*: User:Delta-9

File:Cadenazzo-coat of arms.svg *Source*: http://en.wikipedia.org/w/index.php?title=File:Cadenazzo-coat_of_arms.svg *License*: unknown *Contributors*: User:Delta-9

File:Isone-coat of arms.svg *Source*: http://en.wikipedia.org/w/index.php?title=File:Isone-coat_of_arms.svg *License*: unknown *Contributors*: User:Delta-9

File:Karte Gemeinden der Schweiz 2012.png *Source*: http://en.wikipedia.org/w/index.php?title=File:Karte_Gemeinden_der_Schweiz_2012.png *License*: unknown *Contributors*: Tschubby

file:Sementina-coat of arms.svg *Source*: http://en.wikipedia.org/w/index.php?title=File:Sementina-coat_of_arms.svg *License*: unknown *Contributors*: User:Delta-9

File:Castelgrande Bellinzona.JPG *Source*: http://en.wikipedia.org/w/index.php?title=File:Castelgrande_Bellinzona.JPG *License*: unknown *Contributors*: User:H005

file:Snake-coat of arms.svg *Source*: http://en.wikipedia.org/w/index.php?title=File:Snake-coat_of_arms.svg *License*: unknown *Contributors*: User:Delta-9

File:Bellinzona.Wappen.png *Source*: http://en.wikipedia.org/w/index.php?title=File:Bellinzona.Wappen.png *License*: unknown *Contributors*: User:Mattes, User:Micze

File:Coat of arms of the House of Visconti (1395).svg *Source*: http://en.wikipedia.org/w/index.php?title=File:Coat_of_arms_of_the_House_of_Visconti_(1395).svg *License*: unknown *Contributors*: User:F l a n k e r

Image:Bellinzona Stadtmauer.JPG *Source*: http://en.wikipedia.org/w/index.php?title=File:Bellinzona_Stadtmauer.JPG *License*: unknown *Contributors*: Docu, Ies, Jungpionier, LittleJoe, Manoillon, Tobyc75, Torsch, Warburg, Wst, 1 anonymous edits

File:Castelli.JPG *Source*: http://en.wikipedia.org/w/index.php?title=File:Castelli.JPG *License*: unknown *Contributors*: Aliman5040, Dek09, Docu, Gonce, Jungpionier, Massimo Macconi, MrPanyGoff

File:Catastrophe.jpg *Source*: http://en.wikipedia.org/w/index.php?title=File:Catastrophe.jpg *License*: unknown *Contributors*: Basler Nachrichten

File:Picswiss TI-21-25.jpg *Source*: http://en.wikipedia.org/w/index.php?title=File:Picswiss_TI-21-25.jpg *License*: unknown *Contributors*: Citypeek, Dake, Docu, Roland zh, Testra

File:Picswiss TI-21-34.jpg *Source*: http://en.wikipedia.org/w/index.php?title=File:Picswiss_TI-21-34.jpg *License*: unknown *Contributors*: Dake, Docu, Friedrichstrasse, G.dallorto, Roland zh, Testra, 1 anonymous edits

File:Bellinzona Fernmeldedirektion detail.jpg *Source*: http://en.wikipedia.org/w/index.php?title=File:Bellinzona_Fernmeldedirektion_detail.jpg *License*: unknown *Contributors*: User:Ikiwaner, User:Ultimate Destiny

File:Picswiss TI-21-02.jpg *Source*: http://en.wikipedia.org/w/index.php?title=File:Picswiss_TI-21-02.jpg *License*: unknown *Contributors*: ++gardenfriend++, Dake, Fransvannes, 1 anonymous edits

File:Picswiss TI-21-06.jpg *Source*: http://en.wikipedia.org/w/index.php?title=File:Picswiss_TI-21-06.jpg *License*: unknown *Contributors*: Dake, Roland zh

File:Bellinzona 066.jpg *Source*: http://en.wikipedia.org/w/index.php?title=File:Bellinzona_066.jpg *License*: unknown *Contributors*: User:Aliman5040

File:Bellinzona Montebello.JPG *Source*: http://en.wikipedia.org/w/index.php?title=File:Bellinzona_Montebello.JPG *License*: unknown *Contributors*: Berger (Peter Berger) at de.wikipedia

File:Bellinzona Castello di Sasso Corbaro Turm.jpg *Source*: http://en.wikipedia.org/w/index.php?title=File:Bellinzona_Castello_di_Sasso_Corbaro_Turm.jpg *License*: unknown *Contributors*: User:Clemensfranz

File:Bellinzona Murata.jpg *Source*: http://en.wikipedia.org/w/index.php?title=File:Bellinzona_Murata.jpg *License*: unknown *Contributors*: Clemensfranz

File:Bellinzona SMGrazie Chiostro.jpg *Source*: http://en.wikipedia.org/w/index.php?title=File:Bellinzona_SMGrazie_Chiostro.jpg *License*: unknown *Contributors*: User:Lauròm

File:Bellinzona Collegiata.JPG *Source*: http://en.wikipedia.org/w/index.php?title=File:Bellinzona_Collegiata.JPG *License*: unknown *Contributors*: Fransvannes, Jungpionier

File:Chiesa di San Biagio - Bellinzona.jpg *Source*: http://en.wikipedia.org/w/index.php?title=File:Chiesa_di_San_Biagio_-_Bellinzona.jpg *License*: unknown *Contributors*: Photo uploaded by User:RicciSpeziari. Photographer: Riccardo Speziari

File:PiazzaCollegiataInverno.jpg *Source*: http://en.wikipedia.org/w/index.php?title=File:PiazzaCollegiataInverno.jpg *License*: unknown *Contributors*: Citypeek, Docu, Ori~, Roland zh, Väsk

File:Joseph Mallord William Turner 005.jpg *Source*: http://en.wikipedia.org/w/index.php?title=File:Joseph_Mallord_William_Turner_005.jpg *License*: unknown *Contributors*: Bukk, Docu, Emijrp, Goldfritha, Jastrow, Johnbod, Neddyseagoon, Roland zh

File:Ramparts Bellinzona.jpg *Source*: http://en.wikipedia.org/w/index.php?title=File:Ramparts_Bellinzona.jpg *License*: unknown *Contributors*: User:Kiranchikkadi

Image:Bellinzona 066.jpg *Source*: http://en.wikipedia.org/w/index.php?title=File:Bellinzona_066.jpg *License*: unknown *Contributors*: User:Aliman5040

Image:ChateauMontebelloBellinzona.jpg *Source*: http://en.wikipedia.org/w/index.php?title=File:ChateauMontebelloBellinzona.jpg *License*: unknown *Contributors*: Aliman5040, Docu, Jungpionier, Kilom691

Image:Bellinzona Castello di Sasso Corbaro aussen.jpg *Source*: http://en.wikipedia.org/w/index.php?title=File:Bellinzona_Castello_di_Sasso_Corbaro_aussen.jpg *License*: unknown *Contributors*: Clemensfranz

File:Flag of the Napoleonic Kingdom of Italy.svg *Source*: http://en.wikipedia.org/w/index.php?title=File:Flag_of_the_Napoleonic_Kingdom_of_Italy.svg *License*: unknown *Contributors*: User:Sodacan

File:Flag of Italy (1861-1946).svg *Source*: http://en.wikipedia.org/w/index.php?title=File:Flag_of_Italy_(1861-1946).svg *License*: unknown *Contributors*: User:F l a n k e r

File:Flag of Kingdom of Lombardy-Venetia.svg *Source*: http://en.wikipedia.org/w/index.php?title=File:Flag_of_Kingdom_of_Lombardy-Venetia.svg *License*: unknown *Contributors*: Cycn, Homo lupus, Mikayé

File:StemmaLombardoveneto.png *Source*: http://en.wikipedia.org/w/index.php?title=File:StemmaLombardoveneto.png *License*: unknown *Contributors*: Albertomos, Iguacu

File:Lombardo_Veneto.svg *Source*: http://en.wikipedia.org/w/index.php?title=File:Lombardo_Veneto.svg *License*: unknown *Contributors*: User:ArnoldPlaton

File:Flag of Italy.svg *Source*: http://en.wikipedia.org/w/index.php?title=File:Flag_of_Italy.svg *License*: unknown *Contributors*: Anomie

File:OeLVK0c.png *Source*: http://en.wikipedia.org/w/index.php?title=File:OeLVK0c.png *License*: unknown *Contributors*: Carl Joseph Meyer

File:Francesco I.jpg *Source*: http://en.wikipedia.org/w/index.php?title=File:Francesco_I.jpg *License*: unknown *Contributors*: Boo-Boo Baroo, Che829, Ecummenic, Franzy89, Jimmy44

File:Ferdinand I; Keizer van Oostenrijk.jpg *Source*: http://en.wikipedia.org/w/index.php?title=File:Ferdinand_I;_Keizer_van_Oostenrijk.jpg *License*: unknown *Contributors*: AndreasPraefcke, Bosquete, Carolus, Darwinius, Ecummenic, Gryffindor, Kresspahl, Kürschner, Mcowkin, Mendaliv, Shakko

File:Franz Joseph of Austria young.jpg *Source*: http://en.wikipedia.org/w/index.php?title=File:Franz_Joseph_of_Austria_young.jpg *License*: unknown *Contributors*: AndreasPraefcke, Caro1409, Ecummenic

File:Fantasmi Giubiasco.jpg *Source*: http://en.wikipedia.org/w/index.php?title=File:Fantasmi_Giubiasco.jpg *License*: unknown *Contributors*: Alfo23

file:Giubiasco-coat of arms.svg *Source*: http://en.wikipedia.org/w/index.php?title=File:Giubiasco-coat_of_arms.svg *License*: unknown *Contributors*: User:Delta-9

File:Historische Karte CH 1560.png *Source*: http://en.wikipedia.org/w/index.php?title=File:Historische_Karte_CH_1560.png *License*: unknown *Contributors*: User:Sidonius

File:Linoleum Giubiasco TI jetzt FORBO.jpg *Source*: http://en.wikipedia.org/w/index.php?title=File:Linoleum_Giubiasco_TI_jetzt_FORBO.jpg *License*: unknown *Contributors*: Peter Forster

File:Mornera 14.11.2008 006.jpg *Source*: http://en.wikipedia.org/w/index.php?title=File:Mornera_14.11.2008_006.jpg *License*: unknown *Contributors*: Claudio Vosti

File:Giubiasco.jpg *Source*: http://en.wikipedia.org/w/index.php?title=File:Giubiasco.jpg *License*: unknown *Contributors*: Luca Mascaro

File:Piano di Magadino.JPG *Source*: http://en.wikipedia.org/w/index.php?title=File:Piano_di_Magadino.JPG *License*: unknown *Contributors*: Original uploader was Simone1986 at it.wikipedia

file:Cadenazzo-coat of arms.svg *Source*: http://en.wikipedia.org/w/index.php?title=File:Cadenazzo-coat_of_arms.svg *License*: unknown *Contributors*: User:Delta-9

MIX
Papier aus verantwortungsvollen Quellen
Paper from responsible sources
FSC® C105338

Printed by Books on Demand GmbH, Norderstedt / Germany